Universitext

Peter Orlik
Volkmar Welker

Algebraic Combinatorics

Lectures at a Summer School
in Nordfjordeid, Norway, June 2003

 Springer

Authors

Peter Orlik
Department of Mathematics
University of Wisconsin-Madison
480 Lincoln Drive
Madison, Wisconsin 53706
USA
E-mail: orlik@math.wisc.edu

Volkmar Welker
Philipps-Universität Marburg
Fachbereich Mathematik und Informatik
Hans Meerwein Strasse, Lahnberge
35032 Marburg, Germany
E-mail: welker@mathematik.uni-marburg.de

Editor

Gunnar Fløystad
Department of Mathematics
University of Bergen
Johs.Brunsgate 12
5008 Bergen, Norway
E-mail: Gunnar.Floystad@mi.uib.no

Mathematics Subject Classification (2000): 13D02, 32S22, 52C35; 05E99, 55N25, 55U15

Library of Congress Control Number: 2006938280

ISBN-10 3-540-68375-5 Springer Berlin Heidelberg New York
ISBN-13 978-3-540-68375-9 Springer Berlin Heidelberg New York

Springer is a part of Springer Science+Business Media
springer.com
© Springer-Verlag Berlin Heidelberg 2007

Typesetting: by the authors and techbooks using a Springer TEX macro package

Cover design: *design & production* GmbH, Heidelberg

Printed on acid-free paper   SPIN: 11931218   46/techbooks   5 4 3 2 1 0

Preface

Each year since 1996 the universities of Bergen, Oslo and Trondheim have organized summer schools in Nordfjordeid in various topics in algebra and related fields. Nordfjordeid is the birthplace of Sophus Lie, and is a village on the western coast of Norway situated among fjords and mountains, with spectacular scenery wherever you go. As such it is a welcome place for both Norwegian and international participants and lecturers. The theme for the summer school in 2003 was Algebraic Combinatorics. The organizing committee consisted of Gunnar Fløystad and Stein Arild Strømme (Bergen), Geir Ellingsrud and Kristian Ranestad (Oslo), and Alexej Rudakov and Sverre Smalø (Trondheim). The summer school was partly financed by NorFa-Nordisk Forskerutdanningsakademi.

With combinatorics reaching into and playing an important part of ever more areas in mathematics, in particular algebra, algebraic combinatorics was a timely theme. The fist lecture series "Hyperplane arrangements" was given by Peter Orlik. He came as a refugee to Norway, eighteen years old, after the insurrection in Hungary in 1956. Despite now having lived more than four decades in the United States, he impressed us by speaking fluent Norwegian without a trace of accent. The second lecture series "Discrete Morse theory and free resolutions" was given by Volkmar Welker. These two topics originate back in the second half of the nineteenth century with simple problems on arrangements of lines in the plane and Hilberts syzygy theorem. Although both are classical themes around which mathematics has centered since, there has in recent years been an influx of completely new insights and ideas, and interest in these fields has surged. An attractive feature of both topics is that they relate heavily both to combinatorics, algebra, and topology and, in the case of arrangements, even to analysis, thus giving a rich taste of mathematics. The third lecture series was "Cluster algebras" by Sergei Fomin. This is a recent topic, of this millennium. It has quickly attracted attention due to it giving new insights into classical mathematics as well as giving us fascinating new algebraic structures to study, relating to combinatorics and discrete geometry. This lecture series is however published elsewhere and so is not included here.

But we are pleased to present the first two lecture series in this volume, the topics of which are so natural, classical and inexhaustible that mathematicians will certainly center around them for years to come.

November 2006,   *Gunnar Fløystad*

Contents

Part II Discrete Morse Theory and Free Resolutions

Part I

Lectures on Arrangements:
Combinatorics and Topology

Peter Orlik

These notes represent the revised and enlarged text of my lectures at Nord-fjordeid, Norway in June 2003. My background made this invitation poignant. I arrived in Norway in December 1956 as a shell-shocked and depressed refugee of the defeated Hungarian revolution. Olaf Hauge accepted me into his home, healed my body and spirit, and taught me Norwegian. He was instrumental in helping me reunite with my parents and remained a cherished friend throughout his life. I enrolled in the Technical University in Trondheim, where I came under the guidance of Haakon Waadeland whose encouragement was essential in my decision to switch from engineering to mathematics. He helped me get into graduate school in the United States and I embarked on a career there. My wife, children, and I spent many summers with my parents in Oslo and once again I had the good fortune of developing a lasting friendship with a colleague. Per Holm's hospitality made me welcome in his mathematics department at every visit.

In Hungary my parents and I survived a Fascist dictatorship, World War II, a Communist dictatorship, and a revolution. We found in Norway not only safety, but also the indescribable joy of freedom and respect for human dignity. Even after my father died in 1997, my mother preferred to stay in Oslo, where she felt at home. Although her death in 2002 severed my strongest tie to Norway, this beautiful country and its generous people are embedded in my heart forever.

These notes are dedicated to the memory of my parents
and to the many Norwegians whose help and kindness enriched our lives.

Introduction

Definition

Let V be a vector space of dimension $\ell \geq 1$ over the field $\mathbb{K}$. An arrangement $\mathcal{A} = \{H_1, \ldots, H_n\}$ is a set of $n \geq 0$ hyperplanes in V. In dimension 1, we consider n points in the real line $\mathbb{R}$ or in the complex line $\mathbb{C}$. We shall see later that these seemingly innocent examples lead to interesting problems. In dimension 2, the Selberg arrangement of five lines is shown below. We shall use this arrangement to illustrate definitions and results in Section 1.11.

Fig. 0.1. A Selberg arrangement

Arrangements occur in many branches of mathematics. In combinatorics, we may consider real arrangements and ask for the number of faces of various dimensions in the decomposition of space [54]. In statistics, ranking patterns of the one-dimensional unfolding model may be counted using arrangements [31]. In probability, random walks on arrangements have surprising relations with card shuffling and file management [9]. In topology, the complement of a complex arrangement is a smooth, connected open manifold and we may ask for its topological invariants [38]. The cohomology ring of the complement is a purely combinatorial object, whose algebraic properties have been studied extensively [53]. Multivariable hypergeometric integrals are defined on complements of arrangements, and their classification leads to local system cohomology groups on the complement [1, 2, 30, 39, 46, 51]. These integrals are in turn related to representation theory of Lie algebras and quantum groups [51]. They satisfy certain differential equations which are the Kniznik-Zamolodchikov equations in a special case, of interest in conformal field theory [51, 52]. More generally, they give rise to Gauss-Manin connections in the moduli space of all arrangements of a fixed combinatorial type [2, 39, 15, 16]. A brief history of the subject up to 1990 appeared in [38], where the reader will also find many basic definitions and properties not recorded here. A more recent account of results on hypergeometric integrals was presented in [39]. Although we borrow freely from these sources, our point of view here is different, and we include several new results currently available only in research papers.

Algebraic Combinatorics

In Chapter 1 we collect some aspects of Algebraic Combinatorics related to hyperplane arrangements. For real arrangements, we consider combinatorial questions in Section 1.1, such as counting the number of faces of various types. We define the intersection poset $L(\mathcal{A})$, the Poincaré polynomial $\pi(\mathcal{A}, t)$, and the β invariant $\beta(\mathcal{A})$. We also discuss two other problems involving real arrangements. In Section 1.2 ranking patterns are defined and interpreted as arrangements problems following recent results of [31]. In Section 1.3 random walks on complements of arrangements are considered and their probability distribution is given following work of Bidigare, Billera, Brown, Diaconis, Hanlon, and Rockmore [6, 9, 7].

In the next sections we present several combinatorial constructions. Although they were originally invented as tools to be used in applications, they have since developed a life of their own. It seems appropriate to present them here as objects of study in algebraic combinatorics and relegate their original motivation to the applications in Chapter 2. In order to make the presentation reasonably self-contained, we borrow two sections from [39]: 1.5 and 1.6. In the latter, there was a mistake. The corrected version presented here is due to Hiroaki Terao.

The Orlik-Solomon algebra $A(\mathcal{A})$ is defined and studied in Section 1.4. It was first introduced in [37] to determine the structure of the cohomology ring of the complement of a complex arrangement, see Section 2.1. The Orlik-Solomon algebra $A(\mathcal{A})$ is the quotient of the exterior algebra on the set of hyperplanes by an ideal determined by the combinatorics of the arrangement. Its Hilbert series equals the Poincaré polynomial of the arrangement. The finite dimensional vector space $A(\mathcal{A})$ has a basis called the **nbc** basis, short for no broken circuits. This basis gives rise to the NBC simplicial complex, whose topology is determined in Section 1.5. In particular, its reduced homology is concentrated in one dimension and has a basis of size $\beta(\mathcal{A})$ constructed from the β**nbc** set.

Let $\boldsymbol{\lambda} = (\lambda_1, \ldots, \lambda_n)$ be a set of complex weights for the hyperplanes. Define $a_{\boldsymbol{\lambda}} = \sum_{i=1}^{n} \lambda_i a_i \in A^1(A)$. Note that $a_{\boldsymbol{\lambda}} a_{\boldsymbol{\lambda}} = 0$ because $A(\mathcal{A})$ is the quotient of an exterior algebra. Thus multiplication with a_λ provides a complex $(A^\bullet(\mathcal{A}), a_\lambda)$

$$0 \to A^0(\mathcal{A}) \xrightarrow{a_\lambda} A^1(\mathcal{A}) \xrightarrow{a_\lambda} \ldots \xrightarrow{a_\lambda} A^r(\mathcal{A}) \to 0.$$

In Section 1.6 we study the combinatorial cohomology groups $H^q(A^\bullet(\mathcal{A}), a_\lambda)$ which depend on λ. Thus it is useful to consider the space $\mathbb{C}^n$ of all possible weights λ. Loci in the space of λ where the ranks of these cohomology groups jump are called the resonance varieties. They provide invariants of the Orlik-Solomon algebras [25].

We may go a step further and treat the weights as variables. Aomoto [1] defined a universal complex whose specialization is the complex $(A^\bullet(\mathcal{A}), a_\lambda)$. Let $\mathbf{y} = \{y_1, \ldots, y_n\}$ be a set of indeterminates in one-to-one correspondence with the hyperplanes of $\mathcal{A}$. Let $R = \mathbb{C}[\mathbf{y}]$ be the polynomial ring in $\mathbf{y}$. Define a graded R-algebra:
$$\mathsf{A}^\bullet = \mathsf{A}^\bullet(\mathcal{A}) = R \otimes_{\mathbb{C}} A^\bullet(\mathcal{A}).$$
Let $a_{\mathbf{y}} = \sum_{i=1}^n y_i \otimes a_i \in \mathsf{A}^1$. The complex $(\mathsf{A}^\bullet(\mathcal{A}), a_{\mathbf{y}})$

$$0 \to \mathsf{A}^0(\mathcal{A}) \xrightarrow{a_{\mathbf{y}}} \mathsf{A}^1(\mathcal{A}) \xrightarrow{a_{\mathbf{y}}} \ldots \xrightarrow{a_{\mathbf{y}}} \mathsf{A}^r(\mathcal{A}) \to 0$$

is called the *Aomoto complex*. We determine its cohomology groups under some mild conditions on the variables $\mathbf{y}$ in Section 1.6. Up to this point we considered a single arrangement, first with fixed and later with variable weights for its hyperplanes.

In the rest of this chapter we consider all arrangements of a given combinatorial type. In Section 1.7 we define two arrangements to have the same combinatorial type if there is an order preserving bijection between their intersection posets. This introduces further variables in the setup. We now consider the coefficients of the linear polynomials which define the hyperplanes as variables. The conditions on this system of variables which assures the order preserving bijection on the respective posets will be determined explicitly by the combinatorics. We define dependent sets $\mathrm{Dep}(\mathcal{T})$ for the combinatorial type $\mathcal{T}$. Terao [50] showed that the type of an arrangement is determined by the sets of independent and dependent collections of $\ell + 1$ element subsets of hyperplanes in $\mathcal{A}_\infty$, the projective closure of $\mathcal{A}$ in $\mathbb{CP}^\ell$. This allows for introduction of a partial order on types by reverse inclusion of dependent sets. The type of the general position arrangement $\mathcal{G}$ whose dependent set is empty is the maximal element. For a pair of types $(\mathcal{T}', \mathcal{T})$ we say that $\mathcal{T}$ covers $\mathcal{T}'$ if there is no realizable type between them in the partial order. Moduli spaces of arrangements of a given combinatorial type are also defined in this section.

Certain endomorphisms of the Aomoto complex, called formal connections, arise in the calculation of the Gauss-Manin connection, discussed in detail in Chapter 2. These formal connections are introduced and studied in a purely algebraic setting in Section 1.8 following [16]. Let $(\mathsf{A}^\bullet(\mathcal{G}), a_{\mathbf{y}})$ be the Aomoto complex of a general position arrangement of n hyperplanes in $\mathbb{C}^\ell$. Given any subset S of hyperplanes, we associate with it an endomorphism $\tilde{\omega}_S$ of $(\mathsf{A}^\bullet(\mathcal{G}), a_\mathbf{y})$ and show that it is a cochain homomorphism. In Section 1.9 we define multiplicities $m_S(\mathcal{T})$ for each subset S and each type $\mathcal{T}$.

The Orlik-Solomon algebra and the Aomoto complex depend only on the combinatorial type, so we may label them accordingly. Define

$$\tilde{\omega}(\mathcal{T}', \mathcal{T}) = \sum_{S \in \text{Dep}(\mathcal{T}', \mathcal{T})} m_S(\mathcal{T}') \tilde{\omega}_S$$

where $\text{Dep}(\mathcal{T}', \mathcal{T}) = \text{Dep}(\mathcal{T}') \setminus \text{Dep}(\mathcal{T})$. In order to prove that the endomorphism $\tilde{\omega}(\mathcal{T}', \mathcal{T})$ induces a cochain map $\omega(\mathcal{T}', \mathcal{T})$ on $\mathsf{A}^\bullet(\mathcal{T})$, the Aomoto complex of type $\mathcal{T}$, we show in Section 1.10 that this endomorphism preserves the subcomplex $\mathsf{I}^\bullet(\mathcal{T})$ of the Aomoto complex $\mathsf{A}^\bullet(\mathcal{G})$ corresponding to the Orlik-Solomon ideal of type $\mathcal{T}$. This provides a commutative diagram:

$$
\begin{array}{ccc}
(\mathsf{A}^\bullet(\mathcal{G}), a_\mathbf{y}) & \longrightarrow & (\mathsf{A}^\bullet(\mathcal{T}), a_\mathbf{y}) \\
\downarrow{\scriptstyle \tilde{\omega}(\mathcal{T}', \mathcal{T})} & & \downarrow{\scriptstyle \omega(\mathcal{T}', \mathcal{T})} \\
(\mathsf{A}^\bullet(\mathcal{G}), a_\mathbf{y}) & \longrightarrow & (\mathsf{A}^\bullet(\mathcal{T}), a_\mathbf{y})
\end{array}
$$

The horizontal maps are explicit surjections provided by the respective **nbc** bases. Given weights $\boldsymbol{\lambda}$, the specialization $\mathbf{y} \mapsto \boldsymbol{\lambda}$ in the chain endomorphism $\omega(\mathcal{T}', \mathcal{T})$ defines chain endomorphisms in the Orlik-Solomon algebra $\omega_{\boldsymbol{\lambda}}^q(\mathcal{T}', \mathcal{T}) : A^q(\mathcal{T}) \to A^q(\mathcal{T})$ for $0 \leq q \leq \ell$. These induce endomorphisms

$$\Omega_{\mathcal{C}}^q(\mathcal{T}', \mathcal{T}) : H^q(A^\bullet(\mathcal{T}), a_{\boldsymbol{\lambda}}) \to H^q(A^\bullet(\mathcal{T}), a_{\boldsymbol{\lambda}})$$

for $0 \leq q \leq \ell$ which we call combinatorial Gauss-Manin connections. Most of the definitions and results of this chapter are illustrated in Section 1.11 using the Selberg arrangement. The chapter closes with a set of exercises.

Applications

Chapter 2 contains applications. Section 2.1 concerns topology. Here we assume that V is a complex vector space. The objects of primary interest are the divisor $N(\mathcal{A}) = \cup_{H \in \mathcal{A}} H$ and the complement $\mathsf{M}(\mathcal{A}) = V - N(\mathcal{A})$. The divisor N has interesting non-isolated singularities and we may ask for a resolution described in [39], and the topology of an associated Milnor fiber treated in detail in [19]. The complement M is a connected, smooth, orientable manifold of real dimension 2ℓ. We may ask for its topological invariants, see [38]. Its cohomology ring is isomorphic to the Orlik-Solomon algebra $A(\mathcal{A})$ defined in Section 1.4. In particular, it is a combinatorial invariant.

The problems discussed in the next sections are from analysis. The theory of multivariable hypergeometric integrals involves arrangements. In the Selberg arrangement of Figure 0.1, we may choose coordinates u_1, u_2 in the plane with origin at the lower left vertex and unit equal to the sides of the square. Then the lines are described as the zero sets of the degree 1 polynomials: $u_1 = 0$, $u_1 = 1$, $u_2 = 0$, $u_2 = 1$, and $u_1 = u_2$. In other words, the arrangement is the zero set of the polynomial

$$Q(\mathcal{A}) = u_1(u_1 - 1)u_2(u_2 - 1)(u_1 - u_2).$$

In the general case, introduce coordinates $u_1, \ldots, u_\ell$ in V and choose a linear polynomial α_i for the hyperplane $H_i \in \mathcal{A}$ so H_i is defined by the vanishing of α_i. The product $Q(\mathcal{A}) = \prod_{i=1}^{n} \alpha_i$ is a defining polynomial for $\mathcal{A}$. It is unique up to a constant. Given $\boldsymbol{\lambda}$, we define a multivalued holomorphic function on M by

$$\Phi(u; \boldsymbol{\lambda}) = \prod_{j=1}^{n} \alpha_j^{\lambda_j}.$$

A generalized hypergeometric integral is of the form

$$\int_\sigma \Phi(u; \boldsymbol{\lambda})\eta$$

where σ is a suitable domain of integration and η is a holomorphic form on M, see [2]. This is the Gauss hypergeometric function when $\ell = 1$, $n = 3$ and $\alpha_1 = u, \alpha_2 = u - 1, \alpha_3 = u - x$. Selberg's integral [47] is another special case:

$$\int_0^1 \cdots \int_0^1 (u_1 \cdots u_\ell)^{x-1}[(1 - u_1) \cdots (1 - u_\ell)]^{y-1}|\Delta(u)|^{2z} \, du_1 \ldots du_\ell$$

where $\Delta(u) = \prod_{i<j}(u_j - u_i)$. Assume that the real parts satisfy

$$\Re x > 0, \ \Re y > 0, \ \Re z > -\min\{1/\ell, \Re x/(\ell-1), \Re y/(\ell-1)\}.$$

The corresponding arrangement consists of the coordinate hyperplanes, their parallels $u_i = 1$, and the diagonals $u_i = u_j$ for $i < j$. It is illustrated for $\ell = 2$ in Figure 0.1.

Hypergeometric integrals occur in the representation theory of Lie algebras and quantum groups [51]. In physics, these hypergeometric integrals form solutions to the Knizhnik-Zamolodchikov differential equations in conformal field theory [51]. The space of integrals is identified with a complex rank one local system cohomology group over M. Associated to $\boldsymbol{\lambda}$ we have a rank one representation $\rho : \pi_1(\mathsf{M}) \to \mathbb{C}^*$ given by $\rho(\gamma_j) = t_j$ where $\mathbf{t} = (t_1, \ldots, t_n) \in (\mathbb{C}^*)^n$ is defined by $t_j = \exp(-2\pi i \lambda_j)$ and γ_j is any meridian loop about the hyperplane H_j of $\mathcal{A}$, and a corresponding rank one local system $\mathcal{L} = \mathcal{L}_\mathbf{t} = \mathcal{L}_\lambda$ on M. The first objective is to calculate the local system cohomology groups $H^q(\mathsf{M}, \mathcal{L})$. One method was described in detail in [39], summarizing most results known at that time. It uses the de Rham complex of rational global differential forms on V with arbitrary poles along the divisor N. In this case the cochain groups are infinite dimensional. We must introduce conditions on the weights in order to reduce the problem to a finite dimensional setting. These are the (STV) conditions of Theorem 2.2.6. We call weights $\mathcal{T}$-nonresonant if they satisfy the (STV) conditions for arrangements of combinatorial type $\mathcal{T}$. In this case $H^q(\mathsf{M}; \mathcal{L}) = 0$ for $q \neq \ell$ and $\dim H^\ell(\mathsf{M}; \mathcal{L}) = \beta(\mathcal{A})$. Not much can be said about other weights using this approach.

In these notes the local system cohomology groups are calculated using stratified Morse theory, following [13, 14, 15, 16]. The main advantage is that the cochain groups are finite dimensional for all weights, so this method provides some information for all weights. Note also that when the cochain groups C^q are finite dimensional vector spaces, there are natural projections $\varphi : C^q \to H^q$ constructed as follows. Let Z^q be the group of cocycles. Choose a basis $z_1, \ldots, z_r$ for Z^q and extend it to a basis of C^q by $w_1, \ldots, w_s$. Define $\varphi(z_i) = [z_i]$ as its cohomology class, and $\varphi(w_i) = 0$.

In Section 2.2 we review the stratified Morse theory construction [11, 13] of a finite cochain complex $(K^\bullet(\mathcal{A}), \Delta^\bullet)$, the cohomology of which is naturally isomorphic to $H^\bullet(\mathsf{M}; \mathcal{L})$. This leads to the construction of the universal complex $(\mathsf{K}^\bullet, \Delta^\bullet(\mathbf{x}))$ for local system cohomology where $\mathsf{K}^q = \Lambda \otimes_{\mathbb{C}} K^q$ and $\Lambda = \mathbb{C}[x_1^{\pm 1}, \ldots, x_n^{\pm 1}]$. The Aomoto complex is chain equivalent to the linearization of the universal complex. In Section 2.3 we define the jumping loci of local system cohomology groups, called characteristic varieties, and show that the resonance varieties are tangent cones of the characteristic varieties.

In Section 2.4 we return to the moduli space of arrangements with a fixed combinatorial type $\mathcal{T}$. Let $\mathsf{B}(\mathcal{T})$ be a smooth, connected component of this moduli space. There is a fiber bundle $\pi : \mathsf{M}(\mathcal{T}) \to \mathsf{B}(\mathcal{T})$ whose fibers, $\pi^{-1}(\mathsf{b}) = \mathsf{M}_\mathsf{b}$, are complements of arrangements $\mathcal{A}_\mathsf{b}$ of type $\mathcal{T}$. Since $\mathsf{B}(\mathcal{T})$ is connected, M_b is diffeomorphic to M. The fiber bundle $\pi : \mathsf{M}(\mathcal{T}) \to \mathsf{B}(\mathcal{T})$ is locally trivial. Consequently, given a local system on the fiber, there is an associated flat vector bundle $\pi^q : \mathbf{H}^q(\mathcal{L}) \to \mathsf{B}(\mathcal{T})$, with fiber $(\pi^q)^{-1}(\mathsf{b}) = H^q(\mathsf{M}_\mathsf{b}; \mathcal{L}_\mathsf{b})$ at $\mathsf{b} \in \mathsf{B}(\mathcal{T})$ for each q, $0 \leq q \leq \ell$. Fixing a basepoint $\mathsf{b} \in \mathsf{B}(\mathcal{T})$, the operation of parallel translation of fibers over curves in $\mathsf{B}(\mathcal{T})$ in the vector bundle $\pi^q : \mathbf{H}^q(\mathcal{L}) \to \mathsf{B}(\mathcal{T})$ provides a complex representation

$$\Psi_{\mathcal{T}}^q : \pi_1(\mathsf{B}(\mathcal{T}), \mathsf{b}) \longrightarrow \mathrm{Aut}_{\mathbb{C}}(H^q(\mathsf{M}_\mathsf{b}; \mathcal{L}_\mathsf{b})).$$

The loops of primary interest are those linking moduli spaces of codimension one degenerations of $\mathcal{T}$. Such a degeneration is a type $\mathcal{T}'$ whose moduli space $\mathsf{B}(\mathcal{T}')$ has codimension one in the closure of $\mathsf{B}(\mathcal{T})$. This is equivalent to the combinatorial condition that $\mathcal{T}$ covers $\mathcal{T}'$.

The Gauss hypergeometric function is defined on the complement of the arrangement of three points in $\mathbb{C}$. It satisfies a second order differential equation which, when converted into a system of two linear differential equations, may be interpreted as a Gauss-Manin connection on the moduli space of arrangements of the same combinatorial type [39]. This idea has been generalized to all arrangements by Aomoto [1] and Gelfand [30]. The connection is obtained by differentiating in the moduli space $\mathsf{B}(\mathcal{T})$. Explicit connection matrices were obtained by Aomoto and Kita [2] for arrangements $\mathcal{G}$ in general position and $\mathcal{G}$-nonresonant weights. Unfortunately, this pioneering work is available only in Japanese. The relevant matrices have been reproduced in [39, 15]. The endomorphisms $\tilde{\omega}_S$ in Section 1.8 were modeled on these matrices. Moduli spaces were formalized by Terao [50], who also calculated Gauss-Manin connection matrices in types $\mathcal{T}$ covered by $\mathcal{G}$ and $\mathcal{T}$-nonresonant weights. His results inspired the idea that all Gauss-Manin connections may be obtained by universal constructions.

In [16] we used an equivalent interpretation of the Gauss-Manin connection as a logarithm of the representation Ψ. This allows for local calculations, valid for all arrangements and all weights, and leads to the following result of [16] presented in Section 2.5:

Let M be the complement of an arrangement of type $\mathcal{T}$ and let $\mathcal{L}$ be the local system on M defined by weights $\boldsymbol{\lambda}$. Suppose $\mathcal{T}$ covers $\mathcal{T}'$. Let $\varphi^q : K^q \twoheadrightarrow H^q(\mathsf{M}, \mathcal{L})$ be the natural projection. Then there is an isomorphism $\tau^q : A^q \to K^q$ so that a Gauss-Manin connection endomorphism $\Omega_{\mathcal{L}}^q(\mathsf{B}(\mathcal{T}'), \mathsf{B}(\mathcal{T}))$ in local system cohomology is determined by the equation

$$\varphi^q \circ \tau^q \circ \omega_{\boldsymbol{\lambda}}^q(\mathcal{T}', \mathcal{T}) = \Omega_{\mathcal{L}}^q(\mathsf{B}(\mathcal{T}'), \mathsf{B}(\mathcal{T})) \circ \varphi^q \circ \tau^q.$$

The section closes with an example of how this equation may be used. We provide additional exercises in Section 2.6.

The level of presentation follows what is customary in lectures: some very easy arguments are included to get the audience interested, while more difficult facts may be quoted from the literature.

Much of this material is drawn from joint work with Dan Cohen [13, 14, 15, 16] and Hiroaki Terao [38, 39, 31]. I am grateful to them for all they have taught me in the course of our collaboration. I also thank the organizers of the Summer School, Gunnar Fløystad and Kristian Ranestad, for the invitation to deliver the lectures, for including these notes in the published series, and for helpful comments on the presentation; Dan Cohen for suggestions on the presentation and for providing examples; Hiroaki Terao for permission to use his revisions in Section 1.6; and Anne Britt Orlik of *WritingBarefoot* for her meticulous reading of the manuscript and patient ruthlessness in correcting it.

Madison, October 23, 2006

1

Algebraic Combinatorics

1.1 Chamber Counting

Basic Constructions

Let V be a vector space of dimension ℓ. Let $\mathcal{A}$ be an arrangement of n hyperplanes in V. Let $L = L(\mathcal{A})$ be the set of nonempty intersections of elements of $\mathcal{A}$. An element $X \in L$ is called an *edge* of $\mathcal{A}$. It is convenient to label the hyperplanes $\mathcal{A} = \{H_1, \ldots, H_n\}$ and $X \in L$ by the hyperplanes containing it. Define a *partial order* on L by $X \le Y \Longleftrightarrow Y \subseteq X$. Note that this is reverse inclusion. We agree that V is the unique minimal element of L.

Define a *rank* function on L by $r(X) = \mathrm{codim}X$. Thus $r(V) = 0$, $r(H) = 1$ for $H \in \mathcal{A}$. The rank of $\mathcal{A}$, $r(\mathcal{A})$, is the maximal number of linearly independent hyperplanes in $\mathcal{A}$. It is also the maximal rank of any element in $L(\mathcal{A})$. We call $\mathcal{A}$ *central* if $\cap_{H \in \mathcal{A}} H \ne \emptyset$, where $T = \cap_{H \in \mathcal{A}} H$ is called the center. The ℓ-arrangement $\mathcal{A}$ is called *essential* if it has an element of rank ℓ. Equivalently, $\mathcal{A}$ contains ℓ linearly independent hyperplanes. The poset $L(\mathcal{A})$ of an arrangement has special properties. We call the hyperplanes the atoms of $L(\mathcal{A})$. Given $X, Y \in L(\mathcal{A})$, we define their *meet* by $X \wedge Y = \cap\{Z \in L \mid X \cup Y \subseteq Z\}$. If $X \cap Y \ne \emptyset$, we define their *join* by $X \vee Y = X \cap Y$. Then:

(1) Every element of $L \setminus \{V\}$ is a join of atoms.

(2) For every $X \in L$ all maximal linearly ordered subsets

$$V = X_0 < X_1 < \ldots < X_p = X$$

have the same cardinality. Thus $L(\mathcal{A})$ is a geometric poset. It follows that maximal elements have the same rank.

(3) If $\mathcal{A}$ is central, then all joins exist, so L is a lattice. For all $X, Y \in L$ the rank function satisfies

$$r(X \wedge Y) + r(X \vee Y) \le r(X) + r(Y).$$

Thus for a central arrangement, $L(\mathcal{A})$ is a geometric lattice.

Let $\mathcal{A}$ be a real arrangement. Then $M(\mathcal{A})$ is a union of open, connected components called *chambers*. Each $X \in L(\mathcal{A})$ is similarly decomposed into relatively open connected subsets called *faces*. Let $\mathcal{F}(\mathcal{A})$ be the set of faces. The support $|F|$ of a face F is its affine linear span. Let $\bar{F}$ denote the closure of F in V. The set $\mathcal{F}(\mathcal{A})$ is partially ordered by reverse inclusion: $P \leq Q$ if $Q \subseteq \bar{P}$. We call $\mathcal{F}(\mathcal{A})$ the *face poset* of $\mathcal{A}$.

Chamber Counting

The problem of determining the number of faces of all dimensions in a real arrangement is still unsolved. A more modest problem is to determine the number of chambers. Even this is fairly difficult, so we start with a special case. An arrangement is in *general position* if every intersection of k hyperplanes has rank k. For $k = \ell$ the intersection is a point, and for $k \geq \ell+1$ it is empty. Write $Ch(\mathcal{A})$ for the set of chambers. It is clear that for $\ell = 1$, $|Ch(\mathcal{A})| = n+1$. For $\ell = 2$ we can sketch the first few cases and count:

n	0	1	2	3	4
$|Ch(\mathcal{A})|$	1	2	4	7	11

These numbers are familiar, they are increasing by $n+1$. L. Schläfli proved the following result before 1900.

Theorem 1.1.1. *Given n hyperplanes in general position in $\mathbb{R}^\ell$, the number of chambers in the complement is*

$$|Ch(\mathcal{A})| = \binom{n}{0} + \binom{n}{1} + \cdots + \binom{n}{\ell}.$$

Proof. This formula holds for all ℓ when $n = 0$ and for all n when $\ell = 1$. We proceed by induction. Suppose the formula holds for all n in dimensions $< \ell$ and in dimension ℓ for all arrangements with $< n$ hyperplanes. Now consider an ℓ-arrangement with n hyperplanes. The first $n - 1$ hyperplanes give

$$A = \binom{n-1}{0} + \binom{n-1}{1} + \cdots + \binom{n-1}{\ell}$$

chambers of dimension ℓ. The n-th hyperplane is divided by the first $n - 1$ into

$$B = \binom{n-1}{0} + \binom{n-1}{1} + \cdots + \binom{n-1}{\ell-1}$$

chambers of dimension $\ell - 1$. Each of these chambers divides an ℓ-dimensional chamber into two new chambers. Thus

$$|Ch(\mathcal{A})| = A + B.$$

A binomial coefficient identity completes the proof.   □

What can we expect for an arbitrary arrangement? In 1889, S. Roberts gave a formula in the plane which accounted for the number of regions lost because of multiple points and the number of regions lost because of parallels. The problem is to describe this by a function. The argument in Theorem 1.1.1 proves the next result for arbitrary arrangements. The *deletion* of a nonempty arrangement $\mathcal{A}$ with respect to $H \in \mathcal{A}$ is $\mathcal{A}' = \mathcal{A} - \{H\}$. The *restriction* is $\mathcal{A}'' = \{H \cap K \mid K \in \mathcal{A}'\}$.

Lemma 1.1.2. *Let $\mathcal{A}, \mathcal{A}', \mathcal{A}''$ be a deletion-restriction triple. Then*

$$|Ch(\mathcal{A})| = |Ch(\mathcal{A}')| + |Ch(\mathcal{A}'')|. \quad \Box$$

We need a function with the same recursion. Möbius invented his number theoretic function in 1832. Let $\mathbb{N} = \{1, 2, \ldots\}$ denote the natural numbers. Define $\mu_{\mathbb{N}} : \mathbb{N} \to \{0, \pm 1\}$ by

$$\mu_{\mathbb{N}}(1) = 1$$
$$\mu_{\mathbb{N}}(p_1 \ldots p_r) = (-1)^r \text{ if } p_1, \ldots, p_r \text{ are distinct primes}$$
$$\mu_{\mathbb{N}}(n) = 0 \qquad \text{otherwise.}$$

Suppose $n > 1$ and write $n = p_1^{a_1} \ldots p_r^{a_r}$, where $p_1, \ldots, p_r$ are distinct primes. Since $r > 0$, we have

$$\sum_{d \mid n} \mu_{\mathbb{N}}(d) = \sum_{k=0}^{r} \binom{r}{k} (-1)^k = 0.$$

The condition $\mu_{\mathbb{N}}(1) = 1$ and this formula determine $\mu_{\mathbb{N}}$ recursively.

The Möbius function was generalized to an arbitrary poset L. It is a two-variable function, but we are interested in a one-variable version. Define $\mu : L \to \mathbb{Z}$ by $\mu(V) = 1$, and for $X > V$ by the recursion $\sum_{Y \leq X} \mu(Y) = 0$.

In order to recover the number theoretic Möbius function from the Möbius function of a poset, recall that there is a natural partial order on the set $\mathbb{N}$ defined by $m \leq n \Leftrightarrow m$ divides n. Let $L(\mathbb{N})$ denote this poset. Its unique minimal element is 1. Let $\mu : L(\mathbb{N}) \to \mathbb{Z}$ denote the Möbius function of this poset. Then $\mu_{\mathbb{N}}(n) = \mu(n)$.

The *Poincaré polynomial* of $\mathcal{A}$ is $\pi(\mathcal{A}, t) = \sum_{X \in L} \mu(X)(-t)^{r(X)}$. This polynomial satisfies the following recursion [38, 2.57]:

$$\pi(\mathcal{A}, t) = \pi(\mathcal{A}', t) + t\pi(\mathcal{A}'', t). \tag{1.1}$$

This formula, Lemma 1.1.2, and the fact that the empty arrangement has one chamber prove Zaslavsky's result:

Theorem 1.1.3 ([54]). *Let $\mathcal{A}$ be a real arrangement. The number of chambers in its complement is given by*

$$|Ch(\mathcal{A})| = \pi(\mathcal{A}, 1). \quad \Box$$

There is another poset invariant of importance in the sequel, called the *beta invariant* of the arrangement $\mathcal{A}$ of rank r:

$$\beta(\mathcal{A}) = (-1)^r \pi(\mathcal{A}, -1).$$

The complement of a real arrangement can have bounded chambers only if the arrangement is essential.

Theorem 1.1.4. *Let $\mathcal{A}$ be an essential real arrangement. The number of bounded chambers in its complement is $\beta(\mathcal{A})$.*   □

The beta invariant satisfies a recursion in deletion-restriction. The distinguished hyperplane H is called a *separator* if $r(\mathcal{A}') < r(\mathcal{A})$: removing H reduces the rank of the arrangement. If H is not a separator, then

$$\beta(\mathcal{A}) = \beta(\mathcal{A}') + \beta(\mathcal{A}''). \tag{1.2}$$

Further Constructions

Here are some additional definitions used later.

(1) Given an edge $X \in L$, define a *subarrangement* $\mathcal{A}_X$ of $\mathcal{A}$ by

$$\mathcal{A}_X = \{H \in \mathcal{A} \mid X \subseteq H\}.$$

Here $\mathcal{A}_V$ is the empty ℓ-arrangement Φ_ℓ and if $X \neq V$, then $\mathcal{A}_X$ has center X in any arrangement.

(2) Define an arrangement $\mathcal{A}^X$ in X by

$$\mathcal{A}^X = \{X \cap H \mid H \in \mathcal{A} \setminus \mathcal{A}_X \text{ and } X \cap H \neq \emptyset\}.$$

We call $\mathcal{A}^X$ the *restriction* of $\mathcal{A}$ to X.

(3) The affine ℓ-arrangement $\mathcal{A}$ gives rise to a central $(\ell+1)$-arrangement $\mathbf{c}\mathcal{A}$, called the *cone* over $\mathcal{A}$. Let $\tilde{Q}$ be the homogenized $Q(\mathcal{A})$ with respect to the new variable u_0. Then $Q(\mathbf{c}\mathcal{A}) = u_0\tilde{Q}$ and $|\mathbf{c}\mathcal{A}| = |\mathcal{A}|+1$. There is a natural embedding of $\mathcal{A}$ in $\mathbf{c}\mathcal{A}$ in the subspace $u_0 = 1$. Note that this embedding does not intersect $\ker u_0 = H_\infty$, the *infinite* hyperplane.

(4) Embed V in projective space $\mathbb{P}^\ell$ and call the complement of V the hyperplane at infinity $\bar{H}_\infty$. Let $\bar{H}$ be the projective closure of H and write $\bar{\mathcal{A}} = \cup_{H \in \mathcal{A}} \bar{H}$. We call $\mathcal{A}_\infty = \bar{\mathcal{A}} \cup \{\bar{H}_\infty\}$ the *projective closure* of $\mathcal{A}$. It is an arrangement in $\mathbb{P}^\ell$. Let $u_0, u_1, \dots, u_\ell$ be projective coordinates in $\mathbb{P}^\ell$ so that $\bar{H}_\infty = \ker u_0$. Then $\bar{H} = \ker \tilde{\alpha}_H$ where tilde denotes the homogenized polynomial. Here $Q(\mathcal{A}_\infty) = u_0\tilde{Q}(\mathcal{A})$ and $|\mathcal{A}_\infty| = |\mathcal{A}| + 1$.

(5) Given a nonempty central $(\ell+1)$-arrangement $\mathcal{C}$, we obtain a projective ℓ-arrangement $\mathbb{P}\mathcal{C}$ by viewing the defining homogeneous polynomial $Q(\mathcal{C})$ as a

polynomial in projective coordinates. It is called the *projective quotient*. Here $|\mathcal{C}| = |\mathbb{P}\mathcal{C}|$.

(6) Given a nonempty central $(\ell + 1)$-arrangement $\mathcal{C}$ and a hyperplane $H \in \mathcal{C}$, we define an affine ℓ-arrangement $\mathbf{d}_H\mathcal{C}$, called the *decone* of $\mathcal{C}$ with respect to H. We construct the projective quotient $\mathbb{P}\mathcal{C}$ and choose coordinates so that $\mathbb{P}H = \ker u_0$ is the hyperplane at infinity. By removing it, we obtain the affine arrangement $\mathbf{d}_H\mathcal{C} = \mathbb{P}\mathcal{C} - \mathbb{P}H$. Note that $Q(\mathbf{d}_H\mathcal{C}) = Q(\mathcal{C})|_{u_0=1}$ and $|\mathbf{d}_H\mathcal{C}| = |\mathcal{C}| - 1$. These constructions are interrelated in the diagram below.

$$
\begin{array}{ccc}
 & \mathbf{c}\mathcal{A} & \\
 & \uparrow \quad \searrow & \\
\mathbf{d}_{H_\infty}\mathbf{c}\mathcal{A} \quad = \quad \mathcal{A} & \longrightarrow \quad \mathcal{A}_\infty = & \mathbb{P}\mathbf{c}\mathcal{A}
\end{array}
$$

Let $\mathcal{C}$ be a central arrangement in V with center $T(\mathcal{C}) = \bigcap_{H \in \mathcal{C}} H \neq \emptyset$. We call $\mathcal{C}$ *decomposable* if there exist nonempty subarrangements $\mathcal{C}_1$ and $\mathcal{C}_2$ so that $\mathcal{C} = \mathcal{C}_1 \cup \mathcal{C}_2$ is a disjoint union, and after a linear coordinate change the defining polynomials for $\mathcal{C}_1$ and $\mathcal{C}_2$ have no common variables. This is equivalent to the existence of two nonempty central arrangements so that $\mathcal{C}$ is their product. It is remarkable that this property is determined by the beta invariant [39].

Theorem 1.1.5. *Let $\mathcal{C}$ be a nonempty central arrangement. Then $\beta(\mathbf{d}\mathcal{C}) \geq 0$, and $\mathcal{C}$ is decomposable if and only if $\beta(\mathbf{d}\mathcal{C}) = 0$.* $\qquad\square$

1.2 Ranking Patterns

The simplest arrangement consists of points on the real line, yet we can ask interesting questions about it. The problem stated here has a long history in psychology, sociology, and marketing. It is called the one-dimensional unfolding model. We include some recent results from [31] in the discussion.

The real line $\mathbb{R}$ is viewed as the underlying continuum of a particular attribute, and ℓ objects $x_j \in \mathbb{R}$ ($1 \leq j \leq \ell$) are placed on it according to the amount of this attribute. We may assume that $x_1 < x_2 < \cdots < x_\ell$. (Since the underlying vector space is one-dimensional and the points are hyperplanes, this notation appears to violate our convention that the vector space has dimension ℓ and the number of hyperplanes is n. In fact, we are just anticipating construction of arrangements in vector spaces whose dimension equals the number of points in the problem.)

Each individual preference is represented by a point $y \in \mathbb{R}$. We say that y provides the ranking $(i_1 i_2 \ldots i_\ell)$ if

$$
|y - x_{i_1}| < |y - x_{i_2}| < \cdots < |y - x_{i_\ell}|.
$$

In order to make this unambiguous, we must assume that the midpoints $m_{ij} = (x_i + x_j)/2$ are distinct. Imagine y moving on $\mathbb{R}$. For $y < x_1$ the ranking is

$(12 \ldots \ell)$. When y passes m_{ij}, the adjacent indices i and j are transposed. After passing $\binom{\ell}{2}$ midpoints, $y > x_\ell$ and the ranking is $(\ell \ldots 21)$. The set of these $\binom{\ell}{2} + 1$ permutations is called the *ranking pattern* of $\mathbf{x} = (x_1, \ldots, x_\ell)$.

The question is this: how many ranking patterns are there for given ℓ? If $\ell = 3$, then there is only one: $(123), (213), (231), (321)$. If $\ell = 4$, then there are two: if $m_{14} < m_{23}$, we get

$$(1234), (2134), (2314), (2341), (3241), (3421), (4321)$$

and if $m_{14} > m_{23}$, we get

$$(1234), (2134), (2314), (3214), (3241), (3421), (4321).$$

Here is a brief description of recent work on the subject [31]. We formalize the definitions. Let $R(\mathbf{x}) = \mathbb{R} - \{m_{ij} \mid 1 \leq i < j \leq \ell\}$. Let P_ℓ denote the set of permutations of ℓ letters. Define the ranking map $\mathcal{R}_{\mathbf{x}} : R(\mathbf{x}) \to P_\ell$ by

$$\mathcal{R}_{\mathbf{x}}(y) = (i_1 i_2 \ldots i_\ell) \iff |y - x_{i_1}| < |y - x_{i_2}| < \cdots < |y - x_{i_\ell}|. \qquad (1.3)$$

The assumption $x_1 < x_2 < \cdots < x_\ell$ naturally leads to an arrangement. Let $H_{ij} = \{(x_1, \ldots, x_\ell) \in \mathbb{R}^\ell \mid x_i = x_j\}$. The arrangement

$$\mathcal{B}_\ell = \{H_{ij} \mid 1 \leq i < j \leq \ell\}$$

is called the *braid arrangement*. It has $\binom{\ell}{2}$ hyperplanes. Let $\mathsf{M}(\mathcal{B}_\ell)$ be its complement. Since the ℓ points are distinct, $\mathbf{x} \in \mathsf{M}(\mathcal{B}_\ell)$. Let $\mathbb{S}_\ell$ denote the symmetric group on ℓ letters. The action of $\sigma \in \mathbb{S}_\ell$ on $\mathbb{R}^\ell$ is defined by

$$\sigma(x_1, \ldots, x_\ell) = (x_{\sigma^{-1}(1)}, \ldots, x_{\sigma^{-1}(\ell)}).$$

It is well known that $\mathbb{S}_\ell$ acts simply transitively on the set of chambers of $\mathsf{M}(\mathcal{B}_\ell)$, $Ch(\mathcal{B}_\ell)$. It follows that $|Ch(\mathcal{B}_\ell)| = \ell!$. Let C_0 denote the chamber where $x_1 < x_2 < \cdots < x_\ell$. Given any chamber $C \in Ch(\mathcal{B}_\ell)$ there exists a unique $\sigma \in \mathbb{S}_\ell$ so that $C = \sigma C_0$.

In order to discuss the ranking patterns of these points, we also assume that the midpoints are distinct. This leads to another set of hyperplanes where $m_{pq} = m_{rs}$ or equivalently, $x_p + x_q = x_r + x_s$. Define the index set

$$I_4 = \{(p, q, r, s) \mid 1 \leq p < q \leq \ell, \ p < r < s \leq \ell, \ \text{distinct } p, q, r, s\}.$$

For $(p, q, r, s) \in I_4$, define the hyperplane

$$H_{pqrs} = \{(x_1, \ldots, x_\ell) \in \mathbb{R}^n \mid x_p + x_q = x_r + x_s\}.$$

There are $3\binom{\ell}{4}$ such hyperplanes. The mid-hyperplane arrangement is

$$\mathcal{A}_\ell = \mathcal{B}_\ell \cup \{H_{pqrs} \mid (p, q, r, s) \in I_4\}.$$

The ranking map (1.3) defines ranking patterns for $\mathbf{x} \in C_0 \cap M(\mathcal{A}_\ell)$. Let $r(\ell)$ denote the number of ranking patterns:

$$r(\ell) = |\{im\mathcal{R}_\mathbf{x} \mid \mathbf{x} \in C_0 \cap M(\mathcal{A}_\ell)\}|.$$

The following result of [31] determines $r(\ell)$:

Theorem 1.2.1. *Let* $\sigma \in \mathbb{S}_\ell$ *and* $\mathbf{x}, \mathbf{x}' \in \sigma C_0 \cap M(\mathcal{A}_\ell)$. *Then* $\mathbf{x}$ *and* $\mathbf{x}'$ *have the same ranking pattern if and only if* $\mathbf{x}$ *and* $\mathbf{x}'$ *lie in the same chamber of* $\mathcal{A}_\ell$. *Thus*

$$r(\ell) = \frac{|Ch(\mathcal{A}_\ell)|}{|Ch(\mathcal{B}_\ell)|} = \frac{\pi(\mathcal{A}_\ell, 1)}{\ell!}. \qquad \square$$

This is a satisfactory answer for a mathematician, but explicit knowledge of the values of $r(\ell)$ would be even better. The next table provides these values for $\ell \leq 7$.

ℓ	$\pi(\mathcal{A}_\ell, t)$	$|Ch(\mathcal{A}_\ell)|$	$r(\ell)$
3	$(1+t)(1+2t)$	6	1
4	$(1+t)(1+3t)(1+5t)$	48	2
5	$(1+t)(1+7t)(1+8t)(1+9t)$	1440	12
6	$(1+t)(1+13t)(1+14t)(1+15t)(1+17t)$	120960	168
7	$(1+t)(1+23t)(1+24t)(1+25t)(1+26t)(1+27t)$	23587200	4680

In light of such beautiful factorization of $\pi(\mathcal{A}_\ell, t)$, it is natural to conjecture that factorization into linear terms holds for all ℓ. Unfortunately, we can prove that for $\ell \geq 8$ this is not possible.

1.3 Random Walks

Let $\mathcal{A}$ be a real arrangement and recall $\mathcal{F}(\mathcal{A})$, the face poset of $\mathcal{A}$. There is a particularly efficient way to store the information in the face poset by using the associated oriented matroid. Let $J = \{+, -, 0\}$. We may view each face $F \in \mathcal{F}(\mathcal{A})$ as a map $F : \{1, \ldots, n\} \to J$ defined by $F(k) = \mathrm{sign}\alpha_k(p)$ for any $p \in F$. Note that $F(k) = 0$ if and only if $F \subseteq H_k$, and if $F(k) \neq 0$, then the sign indicates whether F is in the positive or negative half-space determined by H_k. Which side is called *positive* depends on the original choice of α_k. Let $W = J^n$, and let $\pi_k : W \to J$ be the projection onto the k-th coordinate. Define a map $\sigma : \mathcal{F} \to W$ by

$$\pi_k\sigma(F) = \begin{cases} + \text{ if } F(k) > 0, \\ 0 \text{ if } F(k) = 0, \\ - \text{ if } F(k) < 0. \end{cases}$$

Thus the face F gives rise to an n-tuple of elements of J. We illustrate this concept in Figure 1.1, where we labelled the 7 chambers only. There are 9 faces of codimension one and 3 faces of codimension two. Not every n-tuple of elements is in the image of σ. In the example there is no chamber labelled $(-,-,+)$, edge labelled $(0,-,+)$ or vertex labelled $(-,0,0)$.

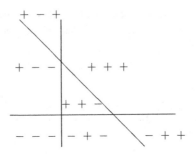

Fig. 1.1. The chambers of $Q(\mathcal{A}) = u_1 u_2 (u_1 + u_2 - 1)$

The face poset is a sharper invariant than the intersection poset. Arrangements $\mathcal{A}$ and $\mathcal{B}$ in Figure 1.2 have isomorphic intersection posets, but they have different face posets.

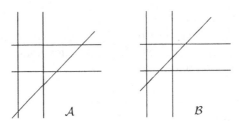

Fig. 1.2. Different face posets

The face poset $\mathcal{F}$ admits the structure of a semigroup. The multiplication is defined for $F, F' \in \mathcal{F}$ by

$$(FF')(k) = \begin{cases} F(k) & \text{if } F(k) \neq 0, \\ F'(k) & \text{if } F(k) = 0. \end{cases}$$

In the special case when C is a chamber, FC is also a chamber for every face F. We may view this as a map $F : Ch(\mathcal{A}) \to Ch(\mathcal{A})$, called the action of F on the chambers. We can define a distance on the set $Ch(\mathcal{A})$. Two chambers are adjacent if they share a codimension one face (wall). A gallery from C to C' is a sequence of chambers $C = C_0, \ldots, C_r = C'$ where successive chambers are adjacent. The distance $d(C, C')$ is the minimal length of a gallery between them. Geometrically, FC is the nearest chamber to C whose closure contains F. We illustrate this in Figure 1.3, where $F = (-,+,0)$ and $C = (+,-,+)$, so $FC = (-,+,+)$.

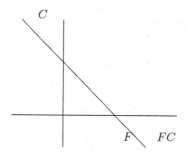

Fig. 1.3. The action of F

Random walks on arrangement complements were first defined using this action by Bidigare, Hanlon, and Rockmore [6]. Our description follows later work by Billera, Brown, and Diaconis [9, 7]. A probability measure w on the face poset assigns to each face $F \in \mathcal{F}$ a nonnegative real number $w(F)$ so that $\sum_{F \in \mathcal{F}} w(F) = 1$. A random walk which starts at the chamber C_0 is the process $(C_r)_{r \geq 0}$ with $C_r = F_r \cdots F_1 C_0$, where $F_1, \ldots, F_r$ are independent and identically distributed choices from w. Equivalently, this is a random walk on the semigroup $\mathcal{F}$ with all states in the ideal of chambers Ch. The transition matrix K of this random walk has entries

$$K(C, C') = \sum_{FC=C'} w(F),$$

representing the probability of moving from C to C' in a single step.

Theorem 1.3.1 ([6, 9]). *Let $\mathcal{A}$ be a real arrangement with intersection poset L and face poset $\mathcal{F}$. Let w be a probability measure on $\mathcal{F}$. Then the transition matrix K is diagonalizable. For each $X \in L$ there is an eigenvalue*

$$\lambda_X = \sum_{\substack{F \in \mathcal{F} \\ F \subseteq X}} w(F)$$

with multiplicity $m_X = (-1)^{r(X)} \mu(X)$. □

The mth power of the matrix K gives the transition probabilities after m steps. Let K_C^m denote the probability distribution of the walk started at C after m steps, so $K_C^m(C') = K^m(C, C')$. A fundamental theorem of Markov chain theory [32, Thm 4.1.4] implies that K_C^m converges to a limit π, as $m \to \infty$ independent of the starting region C. The probability distribution π is called the *stationary* distribution of the walk.

If the measure w is concentrated on a hyperplane, then every walk that enters the hyperplane stays there forever. In this case the stationary distribution is not unique. The measure w is called *separating* if it is not concentrated on any hyperplane of $\mathcal{A}$: for every $H \in \mathcal{A}$ there is a face F with $F \not\subseteq H$ and $w(F) > 0$. The following result determines the stationary distribution and provides an estimate for the rate of convergence to stationarity. For a precise definition of how this is measured, see [7].

Theorem 1.3.2 ([6, 9]). *Let $\mathcal{A}$ be a real arrangement with face poset $\mathcal{F}$. Let w be a probability measure on $\mathcal{F}$ and let K be the transition matrix of the random walk.*

(1) K has a unique stationary distribution π if and only if the measure w is separating.

(2) Assume that w is separating. Then the rate of convergence satisfies

$$\|K_C^m - \pi\| \le \sum_{H \in \mathcal{A}} \lambda_H^m. \quad \square$$

Of the many examples in [6, 9, 7] which illustrate these results, we present the braid arrangement. This arrangement was defined in Section 1.2. Recall that $\mathbb{S}_\ell$ denotes the symmetric group on ℓ letters. Since $\mathbb{S}_\ell$ acts simply transitively on the set of chambers, choice of a distinguished chamber C_0 where $x_1 < x_2 < \cdots < x_\ell$ allows identification of the chambers with the elements of $\mathbb{S}_\ell$: given any chamber C there exists a unique $\sigma \in \mathbb{S}_\ell$ so that $C = \sigma C_0$, so we identify C with σ. The faces can be identified with ordered partitions $B = (B_1, \ldots, B_k)$ of ℓ. For example, $B = (\{2, 5\}, \{1, 4, 7\}, \{3, 6\})$ corresponds to the face defined by $x_2 = x_5 < x_1 = x_4 = x_7 < x_3 = x_6$. The action of a face on a chamber is described in terms of a deck of cards. Suppose C is $\sigma = (\sigma_1, \ldots, \sigma_\ell)$ and a deck of ℓ cards is arranged in the order determined by σ. Let $B = (B_1, \ldots, B_k)$ correspond to the face F. Then FC is a chamber whose permutation is obtained by the operation of B on the cards: remove the cards with labels in B_1 and put them on top without changing their relative order, then remove the cards with labels in B_2 and place them next, and so on. For example if $\sigma = (4215763)$ and $B = (\{2, 5\}, \{1, 4, 7\}, \{3, 6\})$, we get (2541763). Interpretation of the walk as card shuffling provides for interesting examples, depending on the choice of the probability distribution w. Here are two examples.

Random to top

Let $[\ell] = \{1, \ldots, \ell\}$. If we assign a positive weight w_i to each 2-block partition $(\{i\}, \{[\ell] - i\})$ and weight 0 to all other faces, then it corresponds to a scheme called *Tsetlin library* or *random to top* shuffle: a card is repeatedly picked at random according to the weights w_i and is placed on top. This walk represents the situation of a stack of ℓ files, where file i is used with frequency w_i. Each time a file is used, it is replaced on top of the stack. After the process has been running for a long time, the most frequently used files will tend to be near the top.

Riffle shuffle

The braid arrangement is not essential. All faces contain the line $x_1 = \cdots = x_\ell$ corresponding to the single 1-block partition $([\ell])$. The number of 2-block partitions of the form $(s, [\ell] - s)$ where $\emptyset \not\subseteq s \not\subseteq [\ell]$ is $(2^\ell - 2)$. Assign the weight $1/2^\ell$ to each 2-block partition and $2/2^\ell = 1/2^{\ell-1}$ to the 1-block partition. The corresponding shuffling mechanism consists of inverse riffle shuffles. In an ordinary riffle shuffle a deck of cards is divided into two piles which are riffled together. The inverse chooses a set s of cards which are removed (unriffled) and placed on top. Here s can be $\emptyset$ or $[\ell]$, in which case the deck is unchanged. Both of these cases correspond to the action of the 1-block partition. The effect of the choice of weights is that the 2^ℓ subsets $s \subseteq [\ell]$ are equally likely to be unriffled. The stationary distribution π is uniform. For each hyperplane H, $\lambda_H = 1/2$ so the convergence bound from Theorem 1.3.2 gives

$$\|K_\sigma^m - \pi\| \leq \binom{\ell}{2}\left(\frac{1}{2}\right)^m .$$

1.4 The Orlik-Solomon Algebra

In this section we define a combinatorially constructed algebra $A(\mathcal{A})$. It was first introduced in [37] to determine the structure of the cohomology ring of the complement of a complex arrangement, see Section 2.1. It is the quotient of the exterior algebra on the set of hyperplanes by an ideal determined by the combinatorics of the arrangement. Its Hilbert series equals the Poincaré polynomial of the arrangement. The finite dimensional vector space $A(\mathcal{A})$ has a basis called the *nbc* basis, short for no broken circuits. We study the structure of this algebra and how it behaves under deletion and restriction.

Definitions

Let $\mathcal{K}$ be a commutative ring. Let $E^1 = \oplus_{H \in \mathcal{A}} \mathcal{K} e_H$ and let $E = E(\mathcal{A}) = \Lambda(E^1)$ be the exterior algebra of E^1. If $|\mathcal{A}| = n$, then $E = \oplus_{p=0}^n E^p$, where $E^0 = \mathcal{K}$, E^1 agrees with its earlier definition and E^p is spanned over $\mathcal{K}$ by all $e_{H_1} \cdots e_{H_p}$ with $H_k \in \mathcal{A}$. Define a $\mathcal{K}$–linear map $\partial_E = \partial : E \to E$ by $\partial 1 = 0$, $\partial e_H = 1$ and for $p \geq 2$

$$\partial(e_{H_1} \cdots e_{H_p}) = \sum_{k=1}^p (-1)^{k-1} e_{H_1} \cdots \widehat{e_{H_k}} \cdots e_{H_p}$$

for all $H_1, \ldots, H_p \in \mathcal{A}$. If $S = \{H_1, \ldots, H_p\}$, write $e_S = e_{H_1} \cdots e_{H_p}$, $\cap S = H_1 \cap \cdots \cap H_p$, and $|S| = p$. If $p = 0$, we agree that $S = \{\ \}$ is the empty tuple,

$e_S = 1$, and $\cap S = V$. If $\cap S \neq \emptyset$, then we call S *dependent* when $r(\cap S) < |S|$ and *independent* when $r(\cap S) = |S|$. This agrees with linear dependence and independence of the hyperplanes in S.

Let $\mathcal{A}$ be an affine arrangement. Let $I = I(\mathcal{A})$ be the ideal of $E(\mathcal{A})$ generated by

$$\{e_S \mid \cap S = \emptyset\} \cup \{\partial e_S \mid S \text{ is dependent}\}.$$

The *Orlik-Solomon algebra* $A(\mathcal{A})$ is defined by $A(\mathcal{A}) = E(\mathcal{A})/I(\mathcal{A})$. The grading of E induces a grading on $A^\bullet = \oplus_{p=0}^n A^p$.

This algebra was introduced in [37], where we showed that it is isomorphic to the cohomology algebra of the complement $\mathsf{M}(\mathcal{A})$. It has since proved to have other applications as well. We show some basic properties of this algebra established in [37, 38]. This presentation follows the simplified proofs of the excellent survey by Yuzvinsky [53], which also contains a wealth of other topics on the Orlik-Solomon algebra not discussed here.

Properties

Label the hyperplanes $\mathcal{A} = \{H_1, \ldots, H_n\}$ and write $e_i = e_{H_i}$. Let $[n] = \{1, \ldots, n\}$. If $S \subset [n]$ and $i \in S$, then $e_i \partial e_S = \pm e_S$, so I contains all dependent sets. A *circuit* is a minimally dependent set S: S is dependent, but for every $i \in S$, $S_i = S - \{i\}$ is independent. The ideal I is generated by e_S where $\cap S = \emptyset$ and ∂e_S where S is a circuit: if T is any dependent set, then T contains a circuit S. Let $R = T - S$. Then $e_T = \pm e_R e_S$ and $\partial e_T = \pm e_R \partial e_S \pm e_S \partial e_R$. Now $\partial e_S \in I$ by assumption, and $e_S \in I$ because S is dependent. Write $L^q = \{X \in L(\mathcal{A}) \mid r(X) = q\}$.

Proposition 1.4.1. *For $X \in L(\mathcal{A})$, let $E_X(\mathcal{A})$ be the linear span of all e_S with $\cap S = X$, and let $I_X(\mathcal{A}) = I(\mathcal{A}) \cap E_X(\mathcal{A})$. Define $A_X(\mathcal{A}) = E_X(\mathcal{A})/I_X(\mathcal{A})$. For $q \geq 0$*

$$A^q(\mathcal{A}) = \bigoplus_{X \in L^q} A_X^q(\mathcal{A}).$$

Proof. If $\cap S = \emptyset$, then $e_S \in I(\mathcal{A})$. Thus every nonzero element of $A(\mathcal{A})$ is in the projection of some $E_X(\mathcal{A})$. We noted above that the other generators of $I(\mathcal{A})$ are ∂e_S for circuits S. If S is a circuit with $\cap S = X$, then $\cap S_i = X$ for all $i \in S$ and hence $\partial S \in E_X(\mathcal{A})$. Finally, in the direct sum decomposition of $A^q(\mathcal{A})$ now obtained, it is sufficient to sum over $X \in L^q$ since all generators are independent S for which $r(\cap S) = |S|$. $\qquad\square$

Proposition 1.4.2. *For $X \in L(\mathcal{A})$, $\mathcal{A}_X$ is an arrangement, so we may define $E(\mathcal{A}_X)$, $I(\mathcal{A}_X)$, and $A(\mathcal{A}_X)$. Then*

$$I(\mathcal{A}_X) = I(\mathcal{A}) \cap E(\mathcal{A}_X).$$

Proof. Recall that $E(\mathcal{A}_X)$ is generated by e_S where $X \subset \cap S$. The inclusion $I(\mathcal{A}_X) \subset I(\mathcal{A}) \cap E(\mathcal{A}_X)$ is clear. For the converse, suppose $a \in I(\mathcal{A}) \cap E(\mathcal{A}_X)$. Recall that $a \in I(\mathcal{A})$ may be written as

$$a = \sum_{R,S} b_{R,S}\, e_R\, \partial e_S + \sum_{T,U} c_{T,U}\, e_T\, e_U$$

where $b_{R,S}, c_{T,U} \in \mathcal{K}$, S is a circuit, and $\cap U = \emptyset$. Since $\mathcal{A}_X$ is a central arrangement, $U \not\subset \mathcal{A}_X$. If $X \subset \cap S_i$ for some i, then $X \subset \cap S$. Thus a has a summand $a' = \sum_{X \subset \cap(R \cup S)} b_{R,S} e_R\, \partial e_S \in E(\mathcal{A}_X)$ and the generating elements of $a - a'$ are in $\oplus_{Y \neq X} E(\mathcal{A}_Y)$. Thus $a = a' \in I(\mathcal{A}_X)$. $\square$

This result has further consequences for the structure of the Orlik-Solomon algebra:

(1) The homomorphism induced by inclusion $i : A(\mathcal{A}_X) \to A(\mathcal{A})$ is a monomorphism.

(2) If $Y \leq X$, then $A_Y(\mathcal{A}_X) = A_Y(\mathcal{A})$.

Next we show that A is a free $\mathcal{K}$-module and construct its **nbc** basis. Introduce the linear order in $\mathcal{A} = \{H_1, \ldots, H_n\}$ by $H_i \prec H_j$ if $i < j$. Note that it is arbitrary. We call a monomial $e_S \in E(\mathcal{A})$ where $S = (i_1, \ldots, i_p)$ *standard* if $i_1 < \cdots < i_p$. Standard monomials form a multiplicative basis. The linear order in $\mathcal{A}$ induces the degree-lexicographic order on standard monomials. Let $S = (i_1, \ldots, i_p)$ and $T = (j_1, \ldots, j_q)$ be standard. Then $e_S < e_T$ if either $p < q$ or $p = q$ and for some m, $1 \leq m \leq p$, $i_r = j_r$ for $r < m$, and $i_m < j_m$. This order is invariant under multiplication by monomials. Define

$$B(\mathcal{A}) = \{\partial e_S \mid S \text{ is a circuit}\} \cup \{e_T \mid T \text{ is minimal with } \cap T = \emptyset\}.$$

A *broken circuit* is an independent set R such that there exists an index i with the property that (i, R) is a circuit and $i < j$ for all $j \in R$. The initial monomial of ∂e_S is the broken circuit $S_1 = S - \{i_1\}$. The initial monomial of e_T is itself. Let $In(B)$ and $In(I)$ denote the sets of initial monomials. Let $[In(B)]$ and $[In(I)]$ denote the corresponding sets of all monomials divisible by some initial monomial. Let $C = C(\mathcal{A})$ be the linear complement of $[In(B)]$ in $E(\mathcal{A})$, called the **nbc** set, short for no-broken-circuits. This is the set of monomials not divisible by any element of $In(B)$. Let D be the corresponding object for $[In(I)]$. We shall abuse notation by viewing both sets as subsets of $[n]$ and as $\mathcal{K}$-modules. Clearly, $E = I \oplus D$. Thus the surjection $\pi : E \to A$ induces an isomorphism $\pi : D \to A$.

Proposition 1.4.3. *The map* $\pi : C \to A$ *is an isomorphism of $\mathcal{K}$-modules. Thus A is a free $\mathcal{K}$-module and the* **nbc** *set C provides linearly independent generators.*

Proof. Since $[In(B)] \subset [In(I)]$ and hence $D \subset C$, we see that $\pi : C \to A$ is surjective. We must show that **nbc** monomials are independent in A. The $\mathcal{K}$-module C is graded by $[n]$ because it is generated by monomials. It is also graded by $L(\mathcal{A})$ because all its generators are independent sets, and this grading is finer, so $C^p = \oplus_{Y \in L^p} C_Y$ for $0 \le p \le n$. If $e_S \in C$, then each $e_{S_i} \in C$ and hence $\partial e_S \in C$, so $\partial C \subset C$. It follows that $\partial C_X \subset \oplus_{Y < X} C_Y$ where $r(Y) = r(X) - 1$. Next we show that for $X \in L - \{V\}$, $\partial_{|C_X}$ is injective. Let i be the smallest index in $\mathcal{A}_X$. Since elements of C_X contain no broken circuits, $i \in S$ for every $e_S \in C_X$. Thus $e_i C_X = 0$. The general formula $\partial(e_i e_S) = e_S - e_i \partial e_S$ shows that for $e_S \in C_X$, $e_S = e_i \partial e_S$, so ∂ is injective. We complete the proof by induction on the rank. If $r(X) = 0$, then $X = V$, $C_V = \mathcal{K} = A_V$, and $\pi : C_V \to A_V$ is the identity. For $r(X) > 0$ we use the commutative diagram

$$
\begin{array}{ccc}
C_X & \xrightarrow{\partial} & C^{r-1} \\
\downarrow{\scriptstyle \pi} & & \downarrow{\scriptstyle \pi} \\
A_X & \xrightarrow{\partial} & A^{r-1}
\end{array}
$$

We showed that the top map is injective. Since the right vertical map is injective by the induction assumption, so is the left vertical map.   □

It follows from Proposition 1.4.3 that $B(\mathcal{A})$ generates $I(\mathcal{A})$. Note also the following connection between $\mathcal{A}$ and its projective closure $\mathcal{A}_\infty$: S is a circuit in $\mathcal{A}$ if and only if it is a circuit in $\mathcal{A}_\infty$ which does not contain H_∞, and T is minimal with $\cap T = \emptyset$ in $\mathcal{A}$ if and only if (T, H_∞) is a circuit in $\mathcal{A}_\infty$. Thus the circuits of $\mathcal{A}_\infty$ form a generating set for $I(\mathcal{A})$.

Deletion-Restriction

It remains to investigate deletion and restriction. Let $\mathcal{A}$, $\mathcal{A}' = \mathcal{A} - \{H\}$ and $\mathcal{A}'' = \mathcal{A}^H$ be a deletion-restriction triple. We want to construct compatible linear orders. Let $H = H_n$, so the deleted hyperplane is the largest in $\mathcal{A}$. Define $\lambda : \mathcal{A}' \to \mathcal{A}''$ by $\lambda(H) = H \cap H_n$. Next give $\mathcal{A}''$ an arbitrary linear order, $\mathcal{A}'' = \{K_1, \ldots, K_{n''}\}$, where $n'' \le n' = n - 1$. Use this to order $\mathcal{A}'$ as follows: if $\lambda^{-1}(K_1)$ has cardinality k_1, give these hyperplanes the subscripts $1, \ldots, k_1$ in any order; if $\lambda^{-1}(K_2)$ has cardinality k_2, give these hyperplanes the subscripts $k_1 + 1, \ldots, k_1 + k_2$ in any order, and so on. Evidently $k_1 + \cdots + k_{n''} = n'$, so this defines a linear order in $\mathcal{A}'$, and since we have already named H_n, we get a linear order in $\mathcal{A}$. These linear orders satisfy the following properties: (1) the largest hyperplane is deleted; (2) the order in $\mathcal{A}'$ is the restriction of the order in $\mathcal{A}$; and (3) if $\lambda(H) \prec \lambda(K)$ then $H \prec K$. With these orders we may view $\lambda : [n'] \to [n'']$ and extend it to ordered tuples: if $S = (i_1, \ldots, i_p)$, then $\lambda(S) = (\lambda(i_1), \ldots, \lambda(i_p))$. If S is standard, then $\lambda(S)$ is nondecreasing.

Proposition 1.4.4. *Assume $\mathcal{A}$ is nonempty and let $(\mathcal{A}, \mathcal{A}', \mathcal{A}'')$ be a deletion-restriction triple. Then there are exact sequences for $q \ge 0$:*

$$0 \to A^q(\mathcal{A}') \xrightarrow{i} A^q(\mathcal{A}) \xrightarrow{j} A^{q-1}(\mathcal{A}'') \to 0.$$

Proof. Let $i : E' \to E$ be the natural inclusion and define $j : E \to E''$ by $j(e_S) = e_{\lambda(S-\{n\})}$ if $n \in S$, and $j(e_S) = 0$ otherwise. Here i is injective, j is surjective, and $j \circ i = 0$, but $\ker(j) \subset \operatorname{im}(i)$ may not hold. Clearly $i(I') \subset I$. We show that $j(I) \subset I''$. Suppose $S \subset [n]$ and $n \in S$. Write $S' = S - \{n\}$ and note that $\cap S = \cap \lambda(S')$. If $\cap S = \emptyset$, then $\cap \lambda(S') = \emptyset$ and $j(e_S) \in I''$. If S is dependent, then so is $\lambda(S')$, so $j(\partial e_S) = j(\partial(e_{S'}e_n)) = j(\pm e_{S'} + (\partial e_{S'})e_n) = \partial(j(e_S)) \in I''$. Next we argue on the respective **nbc** sets. Since H_n is last in the ordering, $i(C') \subset C$. We show that $j(C) = C''$. Since $j : E \to E''$ is surjective, each standard basis element of E'' is $j(e_S)$ for some S with $n \in S$. The compatible orders imply that such an $S \in C$ if and only if $\lambda(S - \{n\}) \in C''$. In order to show that the sequence

$$0 \to C' \xrightarrow{i} C \xrightarrow{j} C'' \to 0 \tag{1.4}$$

is exact, it suffices to show that $\ker(j) \subset \operatorname{im}(i)$ holds. Since $i(C')$ is generated by e_S with $n \notin S$, it suffices to show that $\ker(j) \cap C$ contains no nonzero linear combination of e_S with $n \in S$. Since $S \in C$ if and only if $\lambda(S - \{n\}) \in C''$, we must prove that λ is injective on the set $S' \subset [n']$ where $(S', n) \in C$. Let $S'_1, S'_2 \subset [n']$ be sets so that $(S'_1, n), (S'_2, n) \in C$ and $\lambda(S'_1) = \lambda(S'_2)$. If $S'_1 \neq S'_2$ then there exist $m \in S'_1$ and $p \in S'_2$ so that $\{m, p, n\}$ is dependent. We may assume $m < p < n$, so (p, n) is a broken circuit. This contradicts $(S'_2, n) \in C$. We conclude that the sequence of **nbc** modules is exact. Proposition 1.4.3 completes the argument.   □

Corollary 1.4.5. *Define the Hilbert series of the graded algebra $A(\mathcal{A})$ by $H(A(\mathcal{A}), t) = \sum \dim(A^q(\mathcal{A}))t^q$. Then $H(A(\mathcal{A}), t) = \pi(\mathcal{A}, t)$.*

Proof. If $\mathcal{A}$ is empty, then $H(A(\mathcal{A}), t) = 1 = \pi(\mathcal{A}, t)$. Let $(\mathcal{A}, \mathcal{A}', \mathcal{A}'')$ be a deletion-restriction triple of a nonempty arrangement. It follows from Proposition 1.4.4 that the Hilbert series satisfies the recursion

$$H(A(\mathcal{A}), t) = H(A(\mathcal{A}'), t) + tH(A(\mathcal{A}''), t).$$

We noted the same recursion for the Poincaré polynomial of the intersection poset in (1.1).   □

1.5 The **NBC** Complex

Definitions

This section is borrowed from [39]. It illustrates a purely combinatorial construction and a mixed combinatorial and topological calculation that is used in Section 1.6 to compute the cohomology groups of a seemingly unrelated complex. The latter is also used in Section 2.2 to calculate the local system cohomology groups of the complement.

Recall the definition of the **nbc** set from the last section. The **nbc** set is closed under taking subsets, hence it forms a simplicial complex called the **nbc** *complex* of $\mathcal{A}$, denoted by NBC. We agree to include the empty set in **nbc** and the empty simplex of dimension -1 in NBC$(\mathcal{A})$. This results in reduced homology and cohomology. If $\mathcal{A}$ has rank r then NBC is a pure $(r-1)$-dimensional complex consisting of independent sets. An $(r-1)$-dimensional simplex of NBC is called an **nbc** frame . A simplex of NBC is *ordered* if its vertices are linearly ordered. We agree to write every element of **nbc** in the standard linear order.

Definition 1.5.1. *Let* $\hat{L} = L \setminus \{V\}$. *Define a map* $\nu : \hat{L} \to \mathcal{A}$ *by*

$$\nu(X) = \min(\mathcal{A}_X).$$

Let $P = (X_1 > \cdots > X_q)$ *be a flag of elements of* $\hat{L}$. *Define*

$$\nu(P) = \{\nu(X_1), \ldots, \nu(X_q)\}.$$

Let $S = \{H_{i_1}, \ldots, H_{i_q}\}$ *be an independent* q-tuple *with* $H_{i_1} \prec \cdots \prec H_{i_q}$. *Define a flag*

$$\xi(S) = (X_1 > \cdots > X_q)$$

of $\hat{L}$, *where* $X_p = \bigcap_{k=p}^{q} H_{i_k}$ *for* $1 \le p \le q$. *A flag* $P = (X_1 > \cdots > X_q)$ *is called an* **nbc** *flag if* $P = \xi(S)$ *for some* $S \in$ **nbc**. *Let* $\xi(\textbf{nbc})$ *denote the set of* **nbc** *flags.*

Lemma 1.5.2. *The maps* ξ *and* ν *induce bijections*

$$\xi : \textbf{nbc} \longrightarrow \xi(\textbf{nbc}) \quad and \quad \nu : \xi(\textbf{nbc}) \longrightarrow \textbf{nbc},$$

which are inverses of each other.

Proof. Let $S = \{H_{i_1}, \ldots, H_{i_q}\} \in$ **nbc**. We show first that $\nu \circ \xi(S) = S$. Suppose $\xi(S) = (X_1 > \cdots > X_q)$. Then $H_{i_p} \supseteq X_p$. If $\min \mathcal{A}_{X_p} \prec H_{i_p}$, then $\{\min \mathcal{A}_{X_p}, H_{i_p}, \ldots, H_{i_q}\}$ is dependent and must contain a circuit. It follows that $\{H_{i_p}, \ldots, H_{i_q}\}$ contains a broken circuit. This contradicts $S \in$ **nbc**. Therefore $\min \mathcal{A}_{X_p} = H_{i_p}$ for $1 \le p \le q$. This implies that $\nu(\xi(S)) = S$, so the map $\xi : \textbf{nbc} \longrightarrow \xi(\textbf{nbc})$ is bijective and $\nu \circ \xi : \textbf{nbc} \longrightarrow \textbf{nbc}$ is the identity map. Thus these maps are inverses of each other. $\square$

Lemma 1.5.3. *We have*

$$\xi(\textbf{nbc}) = \{(X_1 > \cdots > X_q) \mid \nu(X_1) \prec \nu(X_2) \prec \cdots \prec \nu(X_q),$$
$$r(X_p) = q - p + 1 \ (1 \le p \le q)\}.$$

Proof. By Lemma 1.5.2, the left-hand side is contained in the right-hand side. Conversely, let $P = (X_1 > \cdots > X_q)$ belong to the right-hand side. We will show that $\nu(P) \in \mathbf{nbc}$. It is sufficient to derive a contradiction assuming that $\nu(P)$ itself is a broken circuit. There exists $H \in \mathcal{A}$ such that (1) $H \prec \nu(X_1)$; and (2) $\{H\} \cup \nu(P)$ is a circuit. Since

$$X_1 \subseteq \nu(X_1) \cap \cdots \cap \nu(X_q) \subseteq H,$$

we have $H \in \mathcal{A}_{X_1}$ and thus $\nu(X_1) = \min(\mathcal{A}_{X_1}) \preceq H$. This contradicts (1). Thus $\nu(P) \in \mathbf{nbc}$. Since $X_p \subseteq \nu(X_p) \cap \cdots \cap \nu(X_q)$ for $1 \leq p \leq q$, and $r(X_p) = q - p + 1 = r(\nu(X_p) \cap \cdots \cap \nu(X_q))$, we have $P = \xi \circ \nu(P)$, so $P \in \xi(\mathbf{nbc})$. $\qquad\square$

We agreed to delete the last hyperplane $H_n \in \mathcal{A}$ in the deletion-restriction triple $(\mathcal{A}, \mathcal{A}', \mathcal{A}'')$, and we have constructed compatible linear orders in them. Clearly $\nu(K) \prec H_n$ for all $K \in \mathcal{A}''$. Let $\mathbf{nbc}' = \mathbf{nbc}(\mathcal{A}')$, $\mathrm{NBC}' = \mathrm{NBC}(\mathcal{A}')$, $\mathbf{nbc}'' = \mathbf{nbc}(\mathcal{A}'')$, and $\mathrm{NBC}'' = \mathrm{NBC}(\mathcal{A}'')$. Write $\overline{\mathbf{nbc}}'' = \{\{\nu S'', H_n\} \mid S'' \in \mathbf{nbc}''\}$. Equation (1.4) shows that there is a disjoint union

$$\mathbf{nbc} = \mathbf{nbc}' \cup \overline{\mathbf{nbc}}''.$$

Lemma 1.5.4. *Let* $\{X_1, \ldots, X_p\} \subseteq \mathcal{A}''$. *Then*

$$\{X_1, \ldots, X_p\} \in \mathbf{nbc}'' \Longleftrightarrow \{\nu(X_1), \ldots, \nu(X_p), H_n\} \in \mathbf{nbc}.$$

Proof. ($\Rightarrow$): Suppose $\{X_1, \ldots, X_p\} \in \mathbf{nbc}''$. If $\{\nu(X_1), \ldots, \nu(X_p), H_n\}$ contains a broken circuit, then there exists an integer k with $1 \leq k \leq p$ and a hyperplane $H \in \mathcal{A}'$ with $H \prec \nu(X_k)$ such that $\{H, \nu(X_k), \ldots, \nu(X_p), H_n\}$ is linearly dependent. Thus $\{H \cap H_n, X_k, \ldots, X_p\}$ is also linearly dependent and $\nu(H \cap H_n) \preceq H \prec \nu(X_k)$. This implies that $\{X_k, \ldots, X_p\}$ contains a broken circuit, which is a contradiction.

($\Leftarrow$): Suppose $\{\nu(X_1), \ldots, \nu(X_p), H_n\} \in \mathbf{nbc}$. If $\{X_1, \ldots, X_p\}$ contains a broken circuit, then there exists an integer k with $1 \leq k \leq p$ and a hyperplane $X \in \mathcal{A}''$ with $X \prec X_k$ such that $\{X, X_k, \ldots, X_p\}$ is linearly dependent. Thus $\{\nu(X), \nu(X_k), \ldots, \nu(X_p), H_n\}$ is also linearly dependent and $\nu(X) \prec \nu(X_k)$. This implies that $\{\nu(X_k), \ldots, \nu(X_p), H_n\}$ contains a broken circuit, which is a contradiction. $\qquad\square$

Lemma 1.5.5. *If* $\{H_{i_1}, \ldots, H_{i_p}, H_n\} \in \mathbf{nbc}$, *then* $\nu(H_{i_k} \cap H_n) = H_{i_k}$ *for* $1 \leq k \leq p$.

Proof. We may assume $p = 1$ without loss of generality. In general $\nu(H_{i_1} \cap H_n) \preceq H_{i_1}$. If $\nu(H_{i_1} \cap H_n) \prec H_{i_1}$, then $\{\nu(H_{i_1} \cap H_n), H_{i_1}, H_n\}$ is linearly dependent. Thus $\{H_{i_1}, H_n\}$ contains a broken circuit, a contradiction. $\qquad\square$

Homotopy Type

Theorem 1.5.6. *Let $\mathcal{A}$ be an ℓ-arrangement of rank $r = r(\mathcal{A}) \geq 1$. Then* $\mathsf{NBC} = \mathsf{NBC}(\mathcal{A})$ *has the homotopy type of a wedge of spheres,* $\vee_{\beta(\mathcal{A})} S^{r-1}$. *If* $\beta(\mathcal{A}) = 0$, *then* NBC *is contractible.*

Proof. The assertion holds for $r = 1$, since $\beta(\mathcal{A}) = |\mathcal{A}| - 1$ and NBC consists of $|\mathcal{A}|$ points.

If $\mathcal{A}$ is an arrangement with $r \geq 2$, then NBC is path connected. It suffices to show that vertices corresponding to distinct hyperplanes $H_i, H_j \in \mathcal{A}$, $i < j$, are connected. If $X = H_i \cap H_j \neq \emptyset$, then $r(X) = 2 \leq r$. If $H_i = \nu(X)$, then $\{H_i, H_j\}$ is a 1-simplex in NBC. If $\nu(X) \prec H_i$, then $\{\nu(X), H_i\}$ and $\{\nu(X), H_j\}$ are both 1-simplexes in NBC. Thus the vertices H_i and H_j are connected. If $H_i \cap H_j = \emptyset$, then there exists H_k with $H_i \cap H_k \neq \emptyset$ and $H_j \cap H_k \neq \emptyset$. It follows that H_i and H_j are connected via H_k.

If v is a vertex of NBC, then its *star*, $\mathrm{st}(v)$, consists of all open simplexes whose closure contains v. The closure, $\overline{\mathrm{st}(v)}$, is a cone with cone point v. Let $(\mathcal{A}, \mathcal{A}', \mathcal{A}'')$ be a triple with respect to the last hyperplane H_n. Then $\overline{\mathrm{st}(H_n)}$ consists of all simplexes belonging to the set $\{S \in \mathbf{nbc} \mid S \cup \{H_n\} \in \mathbf{nbc}\}$. Also NBC' consists of all simplexes S of NBC with $H_n \notin S$. Thus we have

$$\mathsf{NBC} = \overline{\mathrm{st}(H_n)} \cup \mathsf{NBC}'. \tag{1.5}$$

By Lemma 1.5.4, $\nu\{X_1, \ldots, X_p\} \in \overline{\mathrm{st}(H_n)} \cap \mathsf{NBC}'$. So the map ν induces a simplicial map

$$\nu : \mathsf{NBC}'' \longrightarrow \overline{\mathrm{st}(H_n)} \cap \mathsf{NBC}'.$$

This map is obviously injective. It is also surjective by Lemmas 1.5.5 and 1.5.4. Thus the two simplicial complexes are isomorphic:

$$\mathsf{NBC}'' \simeq \overline{\mathrm{st}(H_n)} \cap \mathsf{NBC}'. \tag{1.6}$$

If $\mathcal{A}$ is an arrangement with $r \geq 3$, then NBC is simply connected. We use induction on $|\mathcal{A}|$. Since $|\mathcal{A}| \geq r$, the induction starts with $|\mathcal{A}| = r$. In this case, $\mathcal{A}$ is isomorphic to the arrangement of the coordinate hyperplanes in r-space. Since any subset of $\mathcal{A}$ is a simplex of NBC, NBC is contractible. Now $\overline{\mathrm{st}(H_n)}$ is a cone with cone point H_n. In particular, it is simply connected. Since $|\mathcal{A}'| < |\mathcal{A}|$, the induction hypothesis implies that NBC' is simply connected. Finally, $r(\mathcal{A}'') = r - 1 \geq 2$, so it follows that NBC'' is path connected. Thus, by (1.5) and (1.6), van Kampen's theorem implies that NBC is simply connected.

Next we want to compute the homology groups of NBC. Integer coefficients are understood. Consider the Mayer–Vietoris sequence for the excisive couple $\{\mathrm{st}(H_n), \mathsf{NBC}'\}$. Using (1.5) and (1.6) we get the long exact sequence

$$\ldots \to H_p(\overline{\mathrm{st}(H_n)}) \oplus H_p(\mathsf{NBC}') \xrightarrow{(i_1, i_2)_*} H_p(\mathsf{NBC})$$

$$\xrightarrow{\partial_*} H_{p-1}(\overline{\mathrm{st}(H_n)} \cap \mathsf{NBC}') \xrightarrow{(j_1, -j_2)_*} H_{p-1}(\overline{\mathrm{st}(H_n)}) \oplus H_{p-1}(\mathsf{NBC}') \to \ldots$$

The fact that $\overline{\mathrm{st}(H_n)}$ is contractible, together with (1.6), gives

$$\ldots \to H_p(\mathsf{NBC'}) \xrightarrow{i_{1*}} H_p(\mathsf{NBC}) \xrightarrow{\partial_*} H_{p-1}(\mathsf{NBC''}) \xrightarrow{j_{1*}} H_{p-1}(\mathsf{NBC'}) \to \ldots$$

(1.7)

Next we show that

$$H_p(\mathsf{NBC}) = \begin{cases} 0 & \text{if } p \neq r-1, \\ \text{free of rank } \beta(\mathcal{A}) & \text{if } p = r-1. \end{cases}$$

We use induction on r, and for fixed r on $|\mathcal{A}|$. We have established the assertion for $r = 1$ and arbitrary $|\mathcal{A}|$. The assertion is also correct for arbitrary r when $|\mathcal{A}| = r$, since in this case we may choose coordinates in V so that $\mathcal{A}$ consists of the coordinate hyperplanes. It follows that $\mathcal{A}$ is a central arrangement and by [38, 2.51] that $\beta(\mathcal{A}) = 0$. For the induction step we assume that the result holds for all arrangements $\mathcal{B}$ with $r(\mathcal{B}) < r$ and for all arrangements $\mathcal{B}$ with $r(\mathcal{B}) = r$ and $|\mathcal{B}| < |\mathcal{A}|$. Consider the exact sequence (1.7). Here we need a case distinction. If H_n is a separator, then $r(\mathcal{A}') < r$. In this case $\mathcal{A}' = \mathcal{A}'' \times \Phi_1$, where Φ_1 is the empty 1-arrangement. Thus $\pi(\mathcal{A}', t) = \pi(\mathcal{A}'', t)$, so (1.1) implies that $\pi(\mathcal{A}, t) = (1+t)\pi(\mathcal{A}'', t)$ and hence $\beta(\mathcal{A}) = 0$. On the other hand, $X \cap H_n \neq \emptyset$ for all $X \in L(\mathcal{A}') \setminus \{V\}$ so $\mathsf{NBC} = \mathrm{st}(H_n)$, which is contractible. If H_n is not a separator, then for $p \neq r-1$ the induction hypothesis implies that $H_p(\mathsf{NBC'}) = H_{p-1}(\mathsf{NBC''}) = 0$ and hence $H_p(\mathsf{NBC}) = 0$. For $p = r-1$, the induction hypothesis implies that $H_{p-1}(\mathsf{NBC''})$ is free of rank $\beta(\mathcal{A}'')$ and $H_p(\mathsf{NBC'})$ is free of rank $\beta(\mathcal{A}')$. The conclusion follows from (1.2).

This allows completion of the proof. For $r = 2$, NBC is 1-dimensional and hence it has the homotopy type of a wedge of circles whose number equals the rank of $H_1(\mathsf{NBC})$. We showed above that this rank is $\beta(\mathcal{A})$. For $r \geq 3$, NBC is simply connected. It follows from the homology calculation and the Hurewicz isomorphism theorem that $\pi_i(\mathsf{NBC}) = 0$ for $1 \leq i < r-1$, and $\pi_{r-1}(\mathsf{NBC}) \simeq H_{r-1}(\mathsf{NBC}; \mathbb{Z})$. The last group is free of rank $\beta(\mathcal{A})$. □

The βnbc Basis

Ziegler [55] defined a subset $\beta\mathbf{nbc}(\mathcal{A})$ of $\mathbf{nbc}(\mathcal{A})$ of cardinality $|\beta\mathbf{nbc}(\mathcal{A})| = \beta(\mathcal{A})$. It has the property that if the simplexes corresponding to $\beta\mathbf{nbc}$ are removed from the complex NBC, the remaining simplicial complex is contractible. The set $\beta\mathbf{nbc}$ is used to construct a basis for the only nontrivial cohomology group $H^{r-1}(\mathsf{NBC})$.

Definition 1.5.7. *A frame B is called a βnbc frame if B is an* **nbc** *frame and for every $H \in B$ there exists $H' \prec H$ in $\mathcal{A}$ such that $(B \setminus \{H\}) \cup \{H'\}$ is a frame. Let $\beta\mathbf{nbc}(\mathcal{A})$ be the set of all βnbc frames. When $\mathcal{A}$ is empty, we agree that $\beta\mathbf{nbc}(\mathcal{A}) = \emptyset$.*

We need to determine what happens to these frames under deletion and restriction. Recall that we have agreed to delete the last hyperplane $H_n \in \mathcal{A}$.

Let $\beta\mathbf{nbc} = \beta\mathbf{nbc}(\mathcal{A})$, $\beta\mathbf{nbc}' = \beta\mathbf{nbc}(\mathcal{A}')$, and $\beta\mathbf{nbc}'' = \beta\mathbf{nbc}(\mathcal{A}'')$. Ziegler [55, Theorem 1.5] proved the following important $\beta\mathbf{nbc}$ recursion:

Theorem 1.5.8. *Write* $\overline{\beta\mathbf{nbc}}'' = \{\{\nu B'', H_n\} \mid B'' \in \beta\mathbf{nbc}''\}$. *If H_n is a separator, then $\beta\mathbf{nbc} = \emptyset$. Otherwise, there is a disjoint union*

$$\beta\mathbf{nbc} = \beta\mathbf{nbc}' \cup \overline{\beta\mathbf{nbc}}''. \quad \square$$

When $\ell = 1$ we agree that $\beta\mathbf{nbc}''$ is empty, so $\overline{\beta\mathbf{nbc}}'' = \{H_n\}$.

For an **nbc** frame $B \in \mathbf{nbc}$ let $B^* \in C^{r-1}(\mathsf{NBC})$ denote the $(r-1)$-cochain dual to B. Thus for an **nbc** frame $B' \in \mathbf{nbc}$, B^* is determined by the formula

$$\langle B^*, B' \rangle = \begin{cases} 1 & \text{if } B' = B \\ 0 & \text{otherwise.} \end{cases}$$

The next result follows from Ziegler's recursion theorem 1.5.8:

Theorem 1.5.9 ([29]). *The set $\{[B^*] \mid B \in \beta\mathbf{nbc}\}$ is a basis for the only nonvanishing cohomology group $H^{r-1}(\mathsf{NBC})$.*

Proof. If H_n is a separator, then $\beta\mathbf{nbc} = \emptyset$. In this case NBC is contractible, as we saw in the proof of Theorem 1.5.6. Suppose that H_n is not a separator. Recall (1.5) and (1.6). The Mayer-Vietoris cohomology exact sequence reduces to the short exact sequence

$$0 \to H^{r-2}(\overline{\mathrm{st}(H_n)} \cap \mathsf{NBC}') \xrightarrow{\partial^*} H^{r-1}(\mathsf{NBC})$$
$$\xrightarrow{(i_1,i_2)^*} H^{r-1}(\overline{\mathrm{st}(H_n)}) \oplus H^{r-1}(\mathsf{NBC}') \to 0$$

because $H^p(\mathsf{NBC}) = 0$, $H^p(\mathsf{NBC}') = 0$, and $H^{p-1}(\mathsf{NBC}'') = 0$ when $p \neq r-1$ by Theorem 1.5.6. We describe the connecting morphism ∂^* explicitly. If B'' is an $(r-2)$-simplex of NBC$''$, then $\nu B''$ is an $(r-2)$-simplex of $\overline{\mathrm{st}(H_n)} \cap \mathsf{NBC}'$ and

$$\{\nu B''\}^* \in C^{r-2}(\overline{\mathrm{st}(H_n)} \cap \mathsf{NBC}').$$

The natural map

$$C^{r-2}(\overline{\mathrm{st}(H_n)}) \oplus C^{r-2}(\mathsf{NBC}') \xrightarrow{(j_1,-j_2)^\#} C^{r-2}(\overline{\mathrm{st}(H_n)} \cap \mathsf{NBC}')$$

sends the element $(\{\nu B''\}^*, 0)$ to $\{\nu B''\}^*$. Note that every $(r-1)$-simplex in $\overline{\mathrm{st}(H_n)}$ includes H_n as a vertex and $\{\nu B'', H_n\}$ is the unique $(r-1)$-simplex in $\overline{\mathrm{st}(H_n)}$ which contains $\nu B''$. Thus the coboundary map

$$\delta : C^{r-2}(\overline{\mathrm{st}(H_n)}) \to C^{r-1}(\overline{\mathrm{st}(H_n)})$$

sends $\{\nu B''\}^*$ to $\{\nu B'', H_n\}^*$. This diagram chase shows that

$$\partial^*([\{\nu B''\}^*]) = [\{\nu B'', H_n\}^*] \in H^{r-1}(\mathsf{NBC}).$$

Now we get the desired result by induction on $|\mathcal{A}|$ using Ziegler's recursion theorem 1.5.8. $\quad \square$

1.6 The Aomoto Complex

This section is borrowed from [39] and contains corrections provided by Hiroaki Terao. Lemma 1.6.2 replaces [39, Lemma 6.1.3] and Corollary 1.6.5 replaces [39, Corollary 6.2.2]. We thank the editors for finding the mistake corrected here. The Aomoto complex will play a prominent role in the applications to local system cohomology and the Gauss-Manin connection in Chapter 2.

Definitions

Let $\lambda = (\lambda_1, \ldots, \lambda_n)$ be a set of complex weights for the hyperplanes and define $a_\lambda = \sum_{i=1}^{n} \lambda_i a_i \in A^1(A)$. Note that $a_\lambda a_\lambda = 0$ because $A(\mathcal{A})$ is a quotient of an exterior algebra. Thus multiplication with a_λ provides a complex $(A^\bullet(\mathcal{A}), a_\lambda)$

$$0 \to A^0(\mathcal{A}) \xrightarrow{a_\lambda} A^1(\mathcal{A}) \xrightarrow{a_\lambda} \ldots \xrightarrow{a_\lambda} A^r(\mathcal{A}) \to 0. \tag{1.8}$$

In this section we study the combinatorial cohomology groups $H^p(A^\bullet(\mathcal{A}), a_\lambda)$.

When we try to use linear algebra to compute these groups, it becomes clear that algebraic relations among the λ_i are of crucial importance.

Aomoto Complex

Aomoto [1] defined a universal complex whose specialization is the complex $(A^\bullet(\mathcal{A}), a_\lambda)$. Let $\mathbf{y} = \{y_1, \ldots, y_n\}$ be a set of indeterminates in one-to-one correspondence with the hyperplanes of $\mathcal{A}$. Let $R = \mathbb{C}[\mathbf{y}]$ be the polynomial ring in $\mathbf{y}$.

Define a graded R-algebra:

$$\mathsf{A}^\bullet = \mathsf{A}^\bullet(\mathcal{A}) = R \otimes_{\mathbb{C}} A^\bullet(\mathcal{A}).$$

Let $a_\mathbf{y} = \sum_{H \in \mathcal{A}} y_H \otimes a_H \in \mathsf{A}^1$. The complex $(\mathsf{A}^\bullet(\mathcal{A}), a_\mathbf{y})$

$$0 \to \mathsf{A}^0(\mathcal{A}) \xrightarrow{a_\mathbf{y}} \mathsf{A}^1(\mathcal{A}) \xrightarrow{a_\mathbf{y}} \ldots \xrightarrow{a_\mathbf{y}} \mathsf{A}^r(\mathcal{A}) \to 0 \tag{1.9}$$

is called the *Aomoto complex*. Let S be a multiplicative closed subset of R. Consider the Aomoto complex of quotients by S

$$0 \to \mathsf{A}_\mathsf{S}^0(\mathcal{A}) \xrightarrow{a_\mathbf{y}} \mathsf{A}_\mathsf{S}^1(\mathcal{A}) \xrightarrow{a_\mathbf{y}} \ldots \xrightarrow{a_\mathbf{y}} \mathsf{A}_\mathsf{S}^r(\mathcal{A}) \to 0, \tag{1.10}$$

where $\mathsf{A}_\mathsf{S}^\bullet = \mathsf{A}_\mathsf{S}^\bullet(\mathcal{A}) = R_\mathsf{S} \otimes_R \mathsf{A}^\bullet(\mathcal{A})$, and R_S is the localization of R at S.

Lemma 1.6.1. *If $\mathcal{C}$ is a nonempty central arrangement and $\mathsf{Y} = \{(\sum_{i=1}^{n} y_i)^m \mid m \geq 0\}$, then the complex $(\mathsf{A}_\mathsf{Y}^\bullet(\mathcal{C}), a_\mathbf{y})$ is acyclic.*

Proof. We showed in Proposition 1.4.3 that $\partial C \subset C$. If $c \in C$, then

$$\partial(a_{\mathbf{y}}c) = (\partial a_{\mathbf{y}})c - a_{\mathbf{y}}(\partial c) = (\sum_{i=1}^{n} y_i)c - a_{\mathbf{y}}(\partial c).$$

Thus ∂ is a contracting chain homotopy.   □

This assertion is false for non-central arrangements. The first equality fails because ∂ is not a derivation in that case. This is easy to check in the arrangement of two points on the line.

Lemma 1.6.2. *Let $\boldsymbol{\lambda}$ be a system of weights. Suppose that S is a multiplicative closed subset of R satisfying $f(\boldsymbol{\lambda}) \neq 0$ whenever $f \in \mathsf{S}$. Denote the evaluation map by $ev_{\boldsymbol{\lambda}} : \mathsf{A}_{\mathsf{S}}^p \to A^p$ defined by $y_i \mapsto \lambda_i$. The evaluation map induces a homomorphism*

$$ev_{\boldsymbol{\lambda}} : H^{\bullet}(\mathsf{A}_{\mathsf{S}}^{\bullet}(\mathcal{A}), a_{\mathbf{y}}) \to H^{\bullet}(A^{\bullet}(\mathcal{A}), a_{\boldsymbol{\lambda}}).$$

(1) The evaluation map $ev_{\boldsymbol{\lambda}} : \mathsf{A}_{\mathsf{S}}^p \to A^p$ is surjective.
(2) Let $r = r(\mathcal{A})$. The evaluation map

$$ev_{\boldsymbol{\lambda}} : H^r(\mathsf{A}_{\mathsf{S}}^{\bullet}(\mathcal{A}), a_{\mathbf{y}}) \to H^r(A^{\bullet}(\mathcal{A}), a_{\boldsymbol{\lambda}})$$

is surjective.

Proof. (1) is obvious. We obtain (2) from (1) since the map involves the top cohomology groups.   □

Next we show that the Aomoto complex $(\mathsf{A}_{\mathsf{D}}^{\bullet}, a_{\mathbf{y}})$ of quotients by a suitable multiplicative closed subset D of R is isomorphic to the cochain complex of the simplicial complex NBC.

Dense Edges

An edge $X \in L(\mathcal{A})$ is called *dense* if and only if the central arrangement $\mathcal{A}_X$ is not decomposable. A combinatorial characterization of dense edges follows from Theorem 1.1.5.

Theorem 1.6.3. *Let $\mathcal{A}$ be an arrangement and let $X \in L(\mathcal{A})$. The following conditions are equivalent:*
(1) X is dense,
(2) $\mathcal{A}_X$ is not decomposable,
(3) $\beta(\mathbf{d}\mathcal{A}_X) \neq 0$,
(4) $\beta(\mathbf{d}\mathcal{A}_X) > 0$.   □

Let $\mathcal{A}_\infty$ be the projective closure of $\mathcal{A}$. We agree that the hyperplane at infinity is largest in the linear order and call it H_{n+1} instead of H_∞.

Define $y_{n+1} = -\sum_{i=1}^n y_i$. Let $D(\mathcal{A}_\infty)$ denote the set of dense edges in $\mathcal{A}_\infty$. The multiplicative subset D of R is generated by

$$\{\sum_{H\in(\mathcal{A}_\infty)_Z} y_H \mid Z \in D(\mathcal{A}_\infty)\}.$$

For a flag $P = (Y_1 > Y_2 > \cdots > Y_q)$ in $\hat{L} = L \setminus \{V\}$, define

$$\Xi_{\mathbf{y}}(P) = \prod_{p=1}^{q} a_{\mathbf{y}}(Y_p) \in \mathsf{A}^q,$$

where $a_{\mathbf{y}}(X) = \sum_{H\in\mathcal{A}_X} y_H \otimes a_H$ for $X \in \hat{L}$. For $S = \{H_{i_1}, \ldots, H_{i_q}\} \in \mathbf{nbc}$, recall that $\xi(S) = (X_1 > \cdots > X_q)$, where $X_p = \bigcap_{k=p}^{q} H_{i_k}$ for $1 \le p \le q$. Let $C^\bullet(\text{NBC}, R)$ be the cochain complex of NBC over R. Note that $C^{-1}(\text{NBC}, R)$ is a rank-one free R-module whose basis is the cochain $\emptyset^*$, dual to $\emptyset$, the only (-1)-simplex. We define

$$\Theta^q : C^{q-1}(\text{NBC}, R) \longrightarrow \mathsf{A}^q \quad (1 \le q \le r)$$

by

$$\Theta^q(\alpha) = \sum_{\substack{S\in\mathbf{nbc} \\ |S|=q}} \alpha(S)\theta_{\mathbf{y}}(S) \in \mathsf{A}^q,$$

where $\alpha \in C^{q-1}(\text{NBC}, R)$ and $\theta_{\mathbf{y}}(S) = \Xi_{\mathbf{y}}(\xi(S))$. For $q = 0$, define

$$\Theta^0 : C^{-1}(\text{NBC}, R) \longrightarrow \mathsf{A}^0$$

by $\Theta^0(\alpha) = \alpha(\emptyset) \in \mathsf{A}^0 = R$. The important result below is due to Schechtman-Varchenko [45] and Brylawski-Varchenko [10]. We state it in a slightly different form using NBC [39] .

Theorem 1.6.4. *The maps $\{\Theta^q\}_{0\le q\le r}$ give a morphism from the (augmented) cochain complex $C^{\bullet-1}(\text{NBC}, R)$ to the Aomoto complex $(\mathsf{A}^\bullet, a_{\mathbf{y}})$. This morphism induces an isomorphism over the ring of quotients R_{D}.*

Proof. Step 1. For the first half, it is sufficient to show that the diagram

$$
\begin{array}{ccc}
C^{q-1}(\text{NBC}, R) & \xrightarrow{\ \delta\ } & C^q(\text{NBC}, R) \\
\Theta^q \downarrow & & \downarrow \Theta^{q+1} \\
\mathsf{A}^q_{\mathbf{y}} & \xrightarrow[a_{\mathbf{y}}]{} & \mathsf{A}^{q+1}_{\mathbf{y}}
\end{array}
$$

is commutative, where δ denotes the coboundary map. It is easy to see that the diagram is commutative when $q = 0$. Let $\emptyset^*$ be the (-1)-cochain dual to the (-1)-simplex $\emptyset$. Then

$$\Theta^1 \circ \delta(\emptyset^*) = \Theta^1\Big(\sum_{H \in \mathcal{A}} \{H\}^*\Big) = \sum_{H \in \mathcal{A}} y_H \otimes a_H = a_{\mathbf{y}} = a_{\mathbf{y}}\Theta^0(\emptyset^*).$$

Suppose $q > 0$. Let $S = \{H_{i_1}, \ldots, H_{i_q}\}$ be a $(q-1)$-simplex in NBC. Let S^* denote the $(q-1)$-cochain dual to S:

$$\langle S^*, S' \rangle = \begin{cases} 1 & \text{if } S' = S, \\ 0 & \text{otherwise.} \end{cases}$$

It suffices to show that

$$\Theta^{q+1} \circ \delta(S^*) = a_{\mathbf{y}}\Theta^q(S^*)$$

by induction on q. Note that

$$\delta(S^*) = \sum_{k=0}^{q}(-1)^k \sum_{I_k}\{H_{i_1}, \ldots, H_{i_k}, H, H_{i_{k+1}}, \ldots, H_{i_q}\}^*,$$

where the second summation is over the set

$$I_k = \{H \in \mathcal{A} \mid \{H_{i_1}, \ldots, H_{i_k}, H, H_{i_{k+1}}, \ldots, H_{i_q}\} \in \mathbf{nbc}\}.$$

Let $\xi(S) = (X_1 > \cdots > X_q)$. It follows from Lemmas 1.5.2 and 1.5.3 that the maps ξ and ν provide a bijection between I_0 and

$$J_0 = \{(Z > X_1 > \ldots X_q) \mid \nu(Z) \prec H_{i_1}, r(Z) = q+1\}$$

and for $1 \le k \le q$ between I_k and

$$J_k = \{(Y_1 > \ldots Y_k > Z > X_{k+1} > \cdots > X_q) \mid H_{i_k} \prec \nu(Z) \prec H_{i_{k+1}},$$
$$r(Z) = q - k + 1, r(Y_j) = q - j + 2, \nu(Y_j) = H_{i_j} \ (1 \le j \le k)\}.$$

Fix Y_1 with $\nu(Y_1) = H_{i_1}$, $r(Y_1) = q+1$, and $Y_1 > X_1$. Define

$$J_1(Y_1) = \{(Z > X_2 > \ldots X_q) \mid \nu(Z) \prec H_{i_2}, r(Z) = q, Y_1 > Z\}$$

and

$$J_k(Y_1) = \{(Y_2 > \ldots Y_k > Z > X_{k+1} > \cdots > X_q) \mid H_{i_k} \prec \nu(Z) \prec H_{i_{k+1}},$$
$$r(Z) = q - k + 1, r(Y_j) = q - j + 2, \nu(Y_j) = H_{i_j} \ (2 \le j \le k)\}$$

for $2 \le k \le q$. Then

$$\sum_{k=1}^{q}(-1)^{k-1}\sum_{P \in J_k}\Xi_{\mathbf{y}}(P)$$

$$= \sum_{\substack{\nu(Y_1) = H_{i_1} \\ r(Y_1) = q+1 \\ Y_1 > X_1}} a_{\mathbf{y}}(Y_1)\left[\sum_{k=1}^{q}(-1)^{k-1}\sum_{P \in J_k(Y_1)}\Xi_{\mathbf{y}}(P) - a_{\mathbf{y}}(X_1)a_{\mathbf{y}}(X_2)\ldots a_{\mathbf{y}}(X_q)\right].$$

By the induction assumption for $\{H_{i_2}, \ldots, H_{i_q}\} \in \mathbf{nbc}(\mathcal{A}_{Y_1})$, we have

$$\sum_{k=1}^{q}(-1)^{k-1} \sum_{P \in J_k(Y_1)} \Xi_{\mathbf{y}}(P) = \Theta^q \circ \delta(\{H_{i_2}, \ldots, H_{i_q}\}^*)$$

$$= a_{\mathbf{y}}(Y_1)\Theta^{q-1}(\{H_{i_2}, \ldots, H_{i_q}\}^*)$$

$$= a_{\mathbf{y}}(Y_1)a_{\mathbf{y}}(X_2)\ldots a_{\mathbf{y}}(X_q).$$

Thus

$$\Theta^{q+1} \circ \delta(S^*) = \sum_{k=0}^{q}(-1)^{k} \sum_{P \in J_k} \Xi_{\mathbf{y}}(P) = \sum_{P \in J_0} \Xi_{\mathbf{y}}(P) - \sum_{k=1}^{q}(-1)^{k-1} \sum_{P \in J_k} \Xi_{\mathbf{y}}(P)$$

$$= \sum_{\substack{\nu(Z) \prec H_{i_1} \\ r(Z) = q+1 \\ Z > X_1}} a_{\mathbf{y}}(Z)a_{\mathbf{y}}(X_1)\ldots a_{\mathbf{y}}(X_q) - \sum_{\substack{\nu(Y_1) = H_{i_1} \\ r(Y_1) = q+1 \\ Y_1 > X_1}} a_{\mathbf{y}}(Y_1)$$

$$\times \left[\sum_{k=1}^{q}(-1)^{k-1} \sum_{P \in J_k(Y_1)} \Xi_{\mathbf{y}}(P) - \sum_{\substack{\nu(Z) = H_{i_1} \\ r(Z) = q \\ Y_1 > Z > X_2}} a_{\mathbf{y}}(Z)a_{\mathbf{y}}(X_2)\ldots a_{\mathbf{y}}(X_q)\right]$$

$$= \sum_{\substack{\nu(Z) \prec H_{i_1} \\ r(Z) = q+1 \\ Z > X_1}} a_{\mathbf{y}}(Z)a_{\mathbf{y}}(X_1)\ldots a_{\mathbf{y}}(X_q)$$

$$- \sum_{\substack{\nu(Y_1) = H_{i_1} \\ r(Y_1) = q+1 \\ Y_1 > X_1}} a_{\mathbf{y}}(Y_1)\left[a_{\mathbf{y}}(Y_1)a_{\mathbf{y}}(X_2)\ldots a_{\mathbf{y}}(X_q) - a_{\mathbf{y}}(X_1)\ldots a_{\mathbf{y}}(X_q)\right]$$

$$= \sum_{\substack{\nu(Z) \prec H_{i_1} \\ r(Z) = q+1 \\ Z > X_1}} a_{\mathbf{y}}(Z)a_{\mathbf{y}}(X_1)\ldots a_{\mathbf{y}}(X_q) + \sum_{\substack{\nu(Y_1) = H_{i_1} \\ r(Y_1) = q+1 \\ Y_1 > X_1}} a_{\mathbf{y}}(Y_1)a_{\mathbf{y}}(X_1)\ldots a_{\mathbf{y}}(X_q)$$

$$= \sum_{\substack{r(Y) = q+1 \\ Y > X_1}} a_{\mathbf{y}}(Y)a_{\mathbf{y}}(X_1)\ldots a_{\mathbf{y}}(X_q) = a_{\mathbf{y}}\Theta^q(S^*).$$

Step 2. We have decompositions

$$C^{q-1}(\mathrm{NBC}, R) = \bigoplus_{\substack{X \in L \\ r(X) = q}} C^{q-1}(\mathrm{NBC}(\mathcal{A}_X), R)$$

and

$$A_{\mathbf{y}}^q = \bigoplus_{\substack{X \in L \\ r(X)=q}} A_{\mathbf{y}}^q(\mathcal{A}_X).$$

Note that the map Θ^q is compatible with these decompositions. In other words, Θ^q induces

$$\Theta_X^q : C^{q-1}(\mathsf{NBC}(\mathcal{A}_X), R) \longrightarrow A_{\mathbf{y}}^q(\mathcal{A}_X)$$

for each $X \in L$ with $r(X) = q$. Since $\mathsf{D}(\mathcal{A}_X) \subseteq \mathsf{D}(\mathcal{A})$, we may assume that $\mathcal{A}$ is nonempty central when we prove the last half of the theorem.

Step 3. Suppose that $\mathcal{A}$ is nonempty central. Let $r = r(\mathcal{A})$. Let $0 \leq q \leq r$. We prove that the induced map

$$\Theta_{\mathsf{D}}^q : C^{q-1}(\mathsf{NBC}, R_{\mathsf{D}}) \longrightarrow A_{\mathsf{D}}^q$$

is an isomorphism by induction on $r \geq 1$. When $r = 1$, Θ_{D}^q $(q = 0, 1)$ are isomorphisms because each y_H is invertible in R_{D}. By Step 2 we have

$$C^{q-1}(\mathsf{NBC}, R_{\mathsf{D}}) = \bigoplus_{\substack{X \in L \\ r(X)=q}} C^{q-1}(\mathsf{NBC}(\mathcal{A}_X), R_{\mathsf{D}})$$

and

$$A_{\mathsf{D}}^q = \bigoplus_{\substack{X \in L \\ r(X)=q}} A_{\mathsf{D}}^q(\mathcal{A}_X).$$

By the induction assumption we may assume that the theorem holds true for $\mathcal{A}_X$ when $r(X) < r$. Thus Θ_{D}^q is an isomorphism for $0 \leq q \leq r-1$. Decompose $\mathcal{A}$ into indecomposable subarrangements:

$$\mathcal{A} = \mathcal{A}_1 \uplus \cdots \uplus \mathcal{A}_m.$$

It is not difficult to see that

$$(A_{\mathsf{D}}^{\bullet}, a_{\mathbf{y}}) \simeq \bigotimes_{i=1}^{m} \left(A_{\mathsf{D}}^{\bullet}(\mathcal{A}_i), \left(\sum_{H \in \mathcal{A}_i} y_H \otimes a_H \right) \right).$$

Note $\mathsf{D}(\mathcal{A}_i) \subseteq \mathsf{D}(\mathcal{A})$ for $1 \leq i \leq m$. Thus the complex $(A_{\mathsf{D}}^{\bullet}, a_{\mathbf{y}})$ is acyclic by Lemma 1.6.1. Since NBC is contractible by Theorem 1.5.6, the (augmented) cochain complex $C^{\bullet-1}(\mathsf{NBC}, R_{\mathsf{D}})$ is also acyclic.

Thus we have a commutative diagram

$$
\begin{array}{ccccccccc}
0 & \longrightarrow & C^{-1}(\mathsf{NBC}, R_{\mathsf{D}}) & \overset{\delta}{\longrightarrow} & \cdots & \overset{\delta}{\longrightarrow} & C^{r-1}(\mathsf{NBC}, R_{\mathsf{D}}) & \longrightarrow & 0 \\
& & \downarrow{\scriptstyle \Theta_{\mathsf{D}}^0} & & & & \downarrow{\scriptstyle \Theta_{\mathsf{D}}^r} & & \\
0 & \longrightarrow & A_{\mathsf{D}}^0 & \underset{a_{\mathbf{y}}}{\longrightarrow} & \cdots & \underset{a_{\mathbf{y}}}{\longrightarrow} & A_{\mathsf{D}}^r & \longrightarrow & 0
\end{array}
$$

whose rows are exact, and the vertical maps $\Theta_{\mathsf{D}}^q (0 \leq q \leq r-1)$ are isomorphisms. Therefore the rightmost vertical map Θ_{D}^r is also an isomorphism. This completes the induction step. $\qquad\square$

Corollary 1.6.5. *Let $\mathcal{A}$ be an arrangement of rank r with projective closure $\mathcal{A}_\infty$. For $X \in L(\mathcal{A}_\infty)$, let $\lambda_X = \sum_{X \subset H_i} \lambda_i$. Assume that $\lambda_X \neq 0$ for every $X \in D(\mathcal{A}_\infty)$. Then*

$$H^p(A^\bullet(\mathcal{A}), a_\lambda) = 0 \text{ for } p \neq r \quad \text{and} \quad \dim H^r(A^\bullet(\mathcal{A}), a_\lambda) = \beta(\mathcal{A}).$$

Proof. We have a commutative diagram

$$
\begin{array}{ccc}
C^{q-1}(\mathrm{NBC}, R_\mathrm{D}) & \xleftarrow{\;j\;} & C^{q-1}(\mathrm{NBC}, \mathbb{C}) \\
\Big\downarrow{\scriptstyle \Theta_\mathrm{D}^q} & & \Big\downarrow{\scriptstyle ev_\lambda \circ \Theta_\mathrm{D}^q \circ j} \\
A_\mathrm{D}^q & \xrightarrow[ev_\lambda]{} & A^q.
\end{array}
$$

Here ev_λ is defined in Lemma 1.6.2 and j is the natural map induced by an extension of the coefficient ring. By Theorem 1.6.4 Θ_D^q is an R_D-isomorphism. The composed map $ev_\lambda \circ \Theta_\mathrm{D}^q \circ j$ is the evaluation of Θ_D^q at $y_H = \lambda_H (H \in \mathcal{A})$. It gives a $\mathbb{C}$-isomorphism: $C^{q-1}(\mathrm{NBC}, \mathbb{C}) \xrightarrow{\sim} A^q$. Since each map in the diagram commutes with the coboundary maps, we get the isomorphism

$$H^q(A^\bullet, a_\lambda) \simeq \tilde{H}^{q-1}(\mathrm{NBC}, \mathbb{C}),$$

where $\tilde{H}$ stands for the reduced cohomology. Theorem 1.5.6 completes the proof. □

The βnbc Basis

Next we find an explicit basis using the β**nbc** set following Falk and Terao [29]. Recall the map Θ^r from Theorem 1.6.4 and the evaluation map ev_λ from Lemma 1.6.2.

$$C^{r-1}(\mathrm{NBC}, R) \xrightarrow{\Theta^r} A^r \xrightarrow{ev_\lambda} A^r.$$

Definition 1.6.6. *Define $\zeta : \beta\mathbf{nbc} \to A^r$ by $\zeta(B) = ev_\lambda \circ \Theta^r(B^*)$. Explicitly, if $B = \{H_{i_1}, \dots, H_{i_r}\}$ is a βnbc frame and $\xi(B) = (X_1 > \cdots > X_r)$ where $X_p = \bigcap_{k=p}^r H_{i_k}$ for $1 \leq p \leq r$, then*

$$\zeta(B) = \prod_{p=1}^r \Big(\sum_{H \in \mathcal{A}_{X_p}} \lambda_H \, a_H \Big).$$

Theorem 1.6.7. *Let $\mathcal{A}$ be an affine arrangement of rank r with projective closure $\mathcal{A}_\infty$. Assume that $\lambda_X \neq 0$ for every $X \in D(\mathcal{A}_\infty)$. Then the set*

$$\{\zeta(B) \in H^r(A, a_\lambda) \mid B \in \beta\mathbf{nbc}\}$$

is a basis for the only nonzero combinatorial cohomology group, $H^r(A, a_\lambda)$.

Proof. We combine the results of Theorem 1.5.9, Theorem 1.6.4, and Lemma 1.6.2(2). □

Resonance Varieties

Each point $\boldsymbol{\lambda} \in \mathbb{C}^n$ gives rise to an element $a_{\boldsymbol{\lambda}} \in A^1$ of the Orlik-Solomon algebra $A = A(\mathcal{A})$. For sufficiently generic $\boldsymbol{\lambda}$ the cohomology $H^q(A^{\bullet}, a_{\boldsymbol{\lambda}})$ vanishes for $q < \ell$ by Corollary 1.6.5. Those $\boldsymbol{\lambda}$ for which $H^q(A^{\bullet}, a_{\boldsymbol{\lambda}})$ does not vanish comprise the *resonance varieties*

$$\mathcal{R}_m^q(A) = \{\boldsymbol{\lambda} \in \mathbb{C}^n \mid \dim H^q(A^{\bullet}, a_{\boldsymbol{\lambda}}) \geq m\}.$$

These subvarieties of $\mathbb{C}^n$ are invariants of the Orlik-Solomon algebra $A^{\bullet}$. See Falk [25] and Libgober-Yuzvinsky [36] for detailed discussions of these varieties. They are clearly contained in $\{\lambda_X = 0 \mid X \in D(\mathcal{A}_{\infty})\}$, which is a union of linear subspaces. Falk [25] conjectured that the resonance varieties themselves are unions of linear subspaces. We return to this question in Section 2.3.

1.7 Combinatorial Types

In the rest of this chapter we move from consideration of a fixed arrangement to the study of all arrangements of a given combinatorial type. Fix a pair (ℓ, n) with $n \geq \ell \geq 1$ and consider families of essential ℓ-arrangements with n linearly ordered hyperplanes. In order to define the notion of combinatorial type and degeneration, we must allow for the coincidence of several hyperplanes. We call these new objects *multi-arrangements* and use *simple* to denote arrangements in the ordinary sense. Since the arrangement is assumed to be essential, we may assume that the ℓ linearly independent hyperplanes are the first ℓ in the linear order. Let $\alpha_i = b_{i,0} + \sum_{j=1}^{\ell} b_{i,j} u_j$ $(i = 1, \ldots, n)$. Note that the coefficient matrix $(b_{i,j})$, $1 \leq i, j \leq \ell$ is invertible by assumption.

We embed the arrangement in projective space as described in Section 1.1 and call the result $\mathcal{A}_{\infty}$. Recall that the hyperplane at infinity is largest in the linear order and we call it H_{n+1} instead of H_{∞}. We may therefore view the projective closure of the arrangement as an $(n+1) \times (\ell+1)$ matrix of complex numbers

$$\mathbf{b} = \begin{pmatrix} b_{1,0} & b_{1,1} & \cdots & b_{1,\ell} \\ b_{2,0} & b_{2,1} & \cdots & b_{2,\ell} \\ \vdots & \vdots & \ddots & \vdots \\ b_{n,0} & b_{n,1} & \cdots & b_{n,\ell} \\ 1 & 0 & \cdots & 0 \end{pmatrix} \tag{1.11}$$

whose rows are the hyperplanes of $\mathcal{A}_{\infty}$. Thus $(\mathbb{CP}^{\ell})^n$ may be viewed as the moduli space of all ordered multi-arrangements in $\mathbb{CP}^{\ell}$ with n hyperplanes together with the hyperplane at infinity.

We call two simple arrangements *combinatorially equivalent* if there is an isomorphism of their posets which preserves the linear order of the hyperplanes. Given an arrangement $\mathcal{A}$, define its dependent sets for $1 < q \leq n+1$,

$$\mathrm{Dep}(\mathcal{A})_q = \{\{j_1, \ldots, j_q\} \mid \mathrm{codim}(H_{j_1} \cap \ldots \cap H_{j_q}) < q\}.$$

Let $\mathrm{Dep}(\mathcal{A}) = \cup_{q \leq \ell+1} \mathrm{Dep}(\mathcal{A})_q$. Two essential simple arrangements are combinatorially equivalent if and only if they have the same dependent sets. We call $\mathcal{T}$ their *combinatorial type* and write $\mathrm{Dep}(\mathcal{T})$. Call a subcollection of $[n+1]$ *realizable* if it is $\mathrm{Dep}(\mathcal{A})$ for a simple arrangement $\mathcal{A}$. Note that an arbitrary subcollection of $[n+1]$ is not necessarily realizable as a dependent set. For example, the collection $\{123, 124, 134\}$ is not realizable as a dependent set, since these dependencies imply the dependence of 234.

Terao [50] proved that the combinatorial type is in fact determined by $\mathrm{Dep}(\mathcal{T})_{\ell+1}$. Given a subset $J \subset [n+1]$ of cardinality $\ell + 1$, let $\Delta_J(\mathbf{b})$ denote the determinant of the $(\ell+1) \times (\ell+1)$ minor of $\mathbf{b}$ whose rows are specified by J. Given a realizable type $\mathcal{T}$, the moduli space of type $\mathcal{T}$ is

$$\mathsf{X}(\mathcal{T}) = \{\mathbf{b} \in (\mathbb{CP}^\ell)^n \mid \Delta_J(\mathbf{b}) = 0 \text{ for } J \in \mathrm{Dep}(\mathcal{T})_{\ell+1}, \ \Delta_J(\mathbf{b}) \neq 0 \text{ else}\}.$$

If $\mathcal{G}$ is the type of a general position arrangement, then $\mathrm{Dep}(\mathcal{G}) = \emptyset$ and the moduli space $\mathsf{X}(\mathcal{G})$ is a dense, open subset of $(\mathbb{CP}^\ell)^n$. We investigate topological properties of this moduli space in Section 2.4.

Define a partial order on combinatorial types as follows: $\mathcal{T} \geq \mathcal{T}' \iff \mathrm{Dep}(\mathcal{T}) \subseteq \mathrm{Dep}(\mathcal{T}')$. The combinatorial type $\mathcal{G}$ is the maximal element with respect to this partial order. Write $\mathcal{T} > \mathcal{T}'$ if $\mathrm{Dep}(\mathcal{T}) \subsetneq \mathrm{Dep}(\mathcal{T}')$. If $\mathcal{T} > \mathcal{T}'$, we say that $\mathcal{T}$ *covers* $\mathcal{T}'$ and $\mathcal{T}'$ is a *degeneration* of $\mathcal{T}$ if there is no realizable combinatorial type $\mathcal{T}''$ with $\mathcal{T} > \mathcal{T}'' > \mathcal{T}'$. In this case we define the relative dependence set

$$\mathrm{Dep}(\mathcal{T}', \mathcal{T}) = \mathrm{Dep}(\mathcal{T}') \setminus \mathrm{Dep}(\mathcal{T}).$$

Terao [50] classified the three codimension-one degeneration types in the moduli space of an arrangement whose only dependent set is the circuit $T = (i_1, \ldots, i_{q+1})$. Let T_p be the q-tuple obtained from T by deleting i_p.

I: Degenerations of $\mathcal{T}$ with $|S \cap T| \leq q - 1$ for all $S \in \mathrm{Dep}(\mathcal{T}', \mathcal{T})$.
II: The collection $\{(T_p, m) \mid m \notin T\}$ for each fixed p, $1 \leq p \leq q + 1$.
III: The collection $\{(T_p, m) \mid 1 \leq p \leq q + 1\}$ for each fixed $m \notin T$.

If $q = 1$, then Type II does not appear. Observe that p denotes a position in the ordered set T while m denotes an element not in T.

1.8 Formal Connections

In the remaining sections of this chapter we define formal connections in the Aomoto complex. These endomorphisms determine combinatorial Gauss-Manin connections and will be used in Chapter 2 to study topological analogs. It will be convenient to refer to the hyperplanes by their subscripts. Let $T = \{i_1, \ldots, i_{q+1}\} \subset \{1, \ldots, n\}$. If order matters, we write $T = (i_1, \ldots, i_{q+1})$ and $a_T = a_{i_1} \cdots a_{i_{q+1}}$. Let $(j, T) = (j, i_1, \ldots, i_{q+1})$ and $T_k = (i_1, \ldots, \hat{i}_k, \ldots, i_{q+1})$.

Let $(\mathsf{A}^\bullet(\mathcal{G}), a_{\mathbf{y}})$ be the Aomoto complex of a general position arrangement of n ordered hyperplanes in $\mathbb{C}^\ell$. We embed the arrangement in projective space as described in Section 1.7 and call the resulting type $\mathcal{G}_\infty$. The fact that the hyperplane at infinity H_{n+1} may be part of a dependent set, but the **nbc** set contains only affine hyperplanes leads to awkward case distinctions which have no geometric significance. We write $S \equiv T$ if S and T are equal sets.

Definition 1.8.1. *Let S be an index set of size $q + 1$. Define an R-linear endomorphism of the Aomoto complex $\tilde{\omega}_S : (\mathsf{A}^\bullet(\mathcal{G}), a_{\mathbf{y}}) \to (\mathsf{A}^\bullet(\mathcal{G}), a_{\mathbf{y}})$ as follows. In the formulas below T is a p-tuple, $n + 1 \notin T$, and $1 \le j \le n$.*
   If $n + 1 \notin S$,

$$\tilde{\omega}_S^p(a_T) = \begin{cases} y_j \partial a_{(j,T)} & \text{if } p = q \text{ and } S \equiv (j, T), \\ a_{\mathbf{y}} \partial a_T & \text{if } p = q + 1 \text{ and } S \equiv T, \\ 0 & \text{otherwise.} \end{cases}$$

   If $n + 1 \in S$,

$$\tilde{\omega}_S^p(a_T) = \begin{cases} -\left(\sum_{j \in [n] - T} y_j\right) a_T & \text{if } p = q \text{ and } S \equiv T \cup \{n+1\}, \\ (-1)^{k-1} y_j a_{(j, T_k)} & \text{if } p = q, \; S \equiv (j, T_k) \cup \{n+1\}, \text{ and } j \notin T, \\ (-1)^k a_{\mathbf{y}} a_{T_k} & \text{if } p = q + 1 \text{ and } S \equiv T_k \cup \{n+1\}, \\ 0 & \text{otherwise.} \end{cases}$$

Proposition 1.8.2. *For every S, the map $\tilde{\omega}_S$ is a cochain homomorphism of the Aomoto complex $(\mathsf{A}^\bullet(\mathcal{G}), a_{\mathbf{y}})$.*

Proof. Let S be an index set of size $q + 1$. Since $\tilde{\omega}_S^p = 0$ for $p \ne q, q + 1$, to show that $\tilde{\omega}_S$ is a cochain map, it suffices to check commutativity in the three squares indicated below.

$$\begin{array}{ccccccccc} \longrightarrow & A^{q-1}(\mathcal{G}) & \xrightarrow{\;a_{\mathbf{y}}\;} & A^q(\mathcal{G}) & \xrightarrow{\;a_{\mathbf{y}}\;} & A^{q+1}(\mathcal{G}) & \xrightarrow{\;a_{\mathbf{y}}\;} & A^{q+2}(\mathcal{G}) & \longrightarrow \\ & \downarrow{\tilde{\omega}_S^{q-1}=0} & & \downarrow{\tilde{\omega}_S^q} & & \downarrow{\tilde{\omega}_S^{q+1}} & & \downarrow{\tilde{\omega}_S^{q+2}=0} & \\ \longrightarrow & A^{q-1}(\mathcal{G}) & \xrightarrow{\;a_{\mathbf{y}}\;} & A^q(\mathcal{G}) & \xrightarrow{\;a_{\mathbf{y}}\;} & A^{q+1}(\mathcal{G}) & \xrightarrow{\;a_{\mathbf{y}}\;} & A^{q+2}(\mathcal{G}) & \longrightarrow \end{array}$$

   If $n + 1 \notin S$, then we may assume that $S = \{1, 2, \ldots, q + 1\}$. Otherwise $S = \{U, n+1\}$ and we may assume that $U = \{n - q + 1, \ldots, n\}$. Since $A^\bullet(\mathcal{G})$ is free on the generators a_T, we may work with these.
   In the first square above we start with $a_T \in A^{q-1}$. Since $\tilde{\omega}_S^{q-1} = 0$, we need to show that $\tilde{\omega}_S^q(a_{\mathbf{y}} a_T) = 0$.
   Suppose $n + 1 \notin S$. Since T must be equivalent to a subset of S, we may assume that $T = (3, 4, \ldots, q + 1)$. Then $\tilde{\omega}_S^q(a_{\mathbf{y}} a_T) = y_1 y_2 (\partial a_{(1,2,T)} + \partial a_{(2,1,T)}) = 0$.
   Suppose $n + 1 \in S$. Since T must be equivalent to a subset of U, we may assume $T = (n - q + 2, \ldots, n)$. Then $a_{\mathbf{y}} a_T = \sum_{j=1}^{n-q+1} y_j a_{(j,T)}$. We get

$$\tilde{\omega}_S^q(a_\mathbf{y} a_T) = \sum_{j=1}^{n-q} y_j \tilde{\omega}_S^q(a_{(j,T)}) + y_{n-q+1} \tilde{\omega}_S^q(a_{(n-q+1,T)})$$

$$= \sum_{j=1}^{n-q} (-1)^2 y_{n-q+1} y_j a_{(n-q+1,j,T)_2} + \sum_{j=1}^{n-q} (-1) y_j y_{n-q+1} a_{(j,n-q+1,T)_1}$$

$$= 0.$$

In the second square we start with $a_T \in \mathsf{A}^q$.

Suppose $n + 1 \notin S$. Since T must be equivalent to a subset of S, we may assume that $T = (1, 2, \ldots, q)$. Then $\tilde{\omega}_S^q(a_T) = y_{q+1} \partial a_{(q+1,T)}$ and $a_\mathbf{y} \tilde{\omega}_S^q(a_T) = y_{q+1} a_\mathbf{y} \partial a_{(q+1,T)}$. The only nonzero term in $\tilde{\omega}_S^{q+1}(a_\mathbf{y} a_T)$ is $\tilde{\omega}_S^{q+1}(y_{q+1} a_{(q+1,T)}) = y_{q+1} a_\mathbf{y} \partial a_{(q+1,T)}$, so the assertion holds.

Suppose $n + 1 \in S$. If $T \equiv U$, then $\tilde{\omega}_S^q(a_T) = (-1)^1 (\sum_{j=1}^{n-q} y_j) a_{(j,T)_1}$ and $a_\mathbf{y} \tilde{\omega}_S^q(a_T) = -(\sum_{j=1}^{n-q} y_j) a_\mathbf{y} a_T$. Also, $a_\mathbf{y} a_T = \sum_{j=1}^{n-q} y_j a_{(j,T)}$, so $\tilde{\omega}_S^{q+1}(a_\mathbf{y} a_T) = -(\sum_{j=1}^{n-q} y_j) a_\mathbf{y} a_T$, as required. If $T \not\equiv U$, then we may assume that $T = (n-q, n-q+1, \ldots, n-1)$. Then $\tilde{\omega}_S^q(a_T) = (-1)^2 y_n a_{(n,T)_2}$, so $a_\mathbf{y} \tilde{\omega}_S^q(a_T) = y_n a_\mathbf{y} a_{(n,T)_2}$. On the other hand, there is only one term in $a_\mathbf{y} a_T$ on which $\tilde{\omega}_S^{q+1}$ is nonzero, namely $y_n a_n a_T$. Since $\tilde{\omega}_S^{q+1}(y_n a_{(n,T)}) = y_n a_\mathbf{y} a_{(n,T)_2}$, the assertion holds.

In the third square we start with $a_T \in \mathsf{A}^{q+1}$. Since $\tilde{\omega}_S^{q+2} = 0$, it suffices to show that $a_\mathbf{y} \tilde{\omega}_S^{q+1}(a_T) = 0$. For any S, this follows from $a_\mathbf{y} a_\mathbf{y} = 0$. $\qquad \square$

Remark 1.8.3. The map $\tilde{\omega}_S^q$ is given by geometric considerations in [2, 50, 39]. There are many possible lifts $\tilde{\omega}_S^{q+1}$ which make $\tilde{\omega}_S$ a cochain map. However, if we require that $\tilde{\omega}_S^{q+1}(a_T) = 0$ unless T is related to S as indicated in the definition, then the lift is unique.

1.9 Multiplicities

The formal connection endomorphisms of the last section were defined for the Aomoto complex of the general position type $\mathcal{G}$. Our aim is to show that they induce endomorphisms for all pairs of types $\mathcal{T}, \mathcal{T}'$ where $\mathcal{T}$ covers $\mathcal{T}'$. Simply adding the endomorphisms $\tilde{\omega}_S$ for all $S \in \mathrm{Dep}(\mathcal{T}', \mathcal{T})$ is insufficient because some degenerations may appear in more than one type. We study these multiplicities next.

Definition 1.9.1. *Given $S \subset [n+1]$, let $N_S(\mathcal{T}) = N_S(\mathbf{b})$ denote the submatrix of (1.11) with rows specified by S. Let $\mathrm{rank}\, N_S(\mathcal{T})$ be the size of the largest minor with nonzero determinant. Define the multiplicity of S in $\mathcal{T}$ by*

$$m_S(\mathcal{T}) = |S| - \mathrm{rank}\, N_S(\mathcal{T}).$$

Let

$$\tilde{\omega}(T) = \sum_{S \in \mathrm{Dep}(T)} m_S(T) \cdot \tilde{\omega}_S \quad \text{and} \quad \tilde{\omega}(T',T) = \sum_{S \in \mathrm{Dep}(T',T)} m_S(T') \cdot \tilde{\omega}_S.$$

The Orlik-Solomon algebra for the combinatorial type $\mathcal{G}$ of general position arrangements is the exterior algebra on n generators truncated at level ℓ. For an arrangement $\mathcal{A}$ of combinatorial type $T \neq \mathcal{G}$, the Orlik-Solomon ideal $I(\mathcal{A})$ depends only on the combinatorial type, so we may write $I(T)$. Thus $A(T) = A(\mathcal{G})/I(T)$. The ideal $I(\mathcal{A})$ gives rise to a subcomplex $\mathsf{I}^\bullet(T)$ of the Aomoto complex $\mathsf{A}^\bullet(\mathcal{G})$, and we have $\mathsf{A}^\bullet(T) = \mathsf{A}^\bullet(\mathcal{G})/\mathsf{I}^\bullet(T)$.

In order to prove that the endomorphisms $\tilde{\omega}(T',T)$ on $\mathsf{A}^\bullet(\mathcal{G})$ induce endomorphisms $\omega(T',T)$ on $\mathsf{A}^\bullet(T)$, we must show that they preserve this subcomplex, $\tilde{\omega}(T',T)(\mathsf{I}^\bullet(T)) \subset \mathsf{I}^\bullet(T)$. This fact is established in Section 1.10.

For the remainder of this section we fix the realizable combinatorial types T, T' and assume that T covers T'. Since the Aomoto complex has dimension ℓ, only $|S| \leq \ell + 1$ can contribute to the maps $\tilde{\omega}(T',T)$. Recall that a generating set for $I(T)$ is obtained from the collection of circuits of $\mathcal{A}_\infty$. Fix a circuit $T \in \mathrm{Dep}(T)_{q+1}$. If $T = (U, n+1)$, then the hyperplanes of T meet at infinity in $\mathcal{A}_\infty$, so the hyperplanes of U have empty intersection in $\mathcal{A}$ and a_U is a generator of $I(T)$. If $n+1 \notin T$, then ∂a_T is a generator of $I(T)$. Let

$$r_T = \begin{cases} a_U & \text{if } T = (U, n+1), \\ \partial a_T & \text{if } n+1 \notin T. \end{cases}$$

It is important to remember that if a circuit T is of size $q+1$, then each element in r_T is a q-tuple. The next observation follows from the definition.

Lemma 1.9.2. *Let T be a $q+1$-circuit and let S be any set. If $|T \cap S| < q-1$, then $\tilde{\omega}_S(r_T) = 0$.*

Lemma 1.9.3. *Let $T \in \mathrm{Dep}(T)_{q+1}$ be a circuit and let $S \in \mathrm{Dep}(T',T)_{q+1}$ be a degeneration of Type I. Then $\tilde{\omega}_S(r_T) = 0$.*

Proof. This follows from Lemma 1.9.2 if $|T \cap S| < q-1$. Suppose $|T \cap S| = q-1$. It follows from the definition of the formal connections that $n+1 \notin T$ and that $n+1 \in S$. In this case, we may assume that $T = (1, \ldots, q+1)$ and $S = \{3, 4, \ldots, q+1, m, n+1\}$ where $m \in [n] \setminus \{T\}$. The only nonzero terms in $\tilde{\omega}_S(\partial a_T)$ are:

$$\tilde{\omega}_S(a_{T_2}) = -y_m a_m a_{3,\ldots,q+1} = \tilde{\omega}_S(a_{T_1}).$$

Since these terms appear with opposite signs in ∂a_T, we conclude that $\tilde{\omega}_S(\partial a_T) = 0$. $\qquad\square$

Let $T \in \mathrm{Dep}(T)_{q+1}$ be a circuit and recall that every degeneration of T is of Type I, II, or III. We refer to $S \in \mathrm{Dep}(T',T)$ as T-*relevant* if $\tilde{\omega}_S(r_T) \neq 0$.

For such S we have $|S| = q$ or $|S| = q+1$. It follows from Lemma 1.9.3 that in all further considerations of T-relevant endomorphisms we may assume that the degenerations are of Types II and III.

Lemma 1.9.4. *Let T be a set of cardinality $q + 1$. If $T_i, T_j \in \mathrm{Dep}(\mathcal{T})$ for $i \neq j$, then $T_i \in \mathrm{Dep}(\mathcal{T})$ for all i, $1 \leq i \leq q + 1$.*

Proof. Without loss, assume that $T_1, T_2 \in \mathrm{Dep}(\mathcal{T})$. Then there are nonzero vectors $\boldsymbol{\alpha} = (0, \alpha_2, \alpha_3, \dots, \alpha_{q+1})$ and $\boldsymbol{\beta} = (\beta_1, 0, \beta_3, \dots, \beta_{q+1})$ which are annihilated by the rows of the matrix (1.11) indexed by T_1 and T_2, respectively. If $\alpha_i = 0$ for some $i \neq 1, 2$, then $\boldsymbol{\alpha}$ is annihilated by the rows corresponding to T_i, hence $T_i \in \mathrm{Dep}(\mathcal{T})$. If $\alpha_i \neq 0$, then $\alpha_i \boldsymbol{\beta} - \beta_i \boldsymbol{\alpha}$ is a nonzero vector annihilated by the rows corresponding to T_i, hence $T_i \in \mathrm{Dep}(\mathcal{T})$. $\square$

Proposition 1.9.5. *Let $T \in \mathrm{Dep}(\mathcal{T})_{q+1}$ be a circuit. Then there is at most one $j \in T$ so that $T_j \in \mathrm{Dep}(\mathcal{T}', \mathcal{T})_q$.*

Proof. By relabeling the hyperplanes we may assume that $T = (1, \dots, q + 1)$. Then $T_i = T \setminus \{i\}$. Since T is a circuit, T_i is independent in type $\mathcal{T}$ for each $i \in T$. Since $\mathcal{T}$ is the type of an arrangement which contains ℓ linearly independent hyperplanes, there exists a set $J \subset [n]$ of cardinality $\ell - q$ so that $T_1 \cup J \cup \{n + 1\}$ is independent in $\mathcal{T}$. We assert that $T_i \cup J \cup \{n + 1\}$ is independent in $\mathcal{T}$ for all $i \in T$.

Suppose otherwise. If, for instance, $T_{q+1} \cup J \cup \{n + 1\} \in \mathrm{Dep}(\mathcal{T})$, then there are constants α_i, β_j, ξ, not all zero, so that

$$\sum_{i=1}^{q} \alpha_i \mathbf{b}_i + \sum_{j \in J} \beta_j \mathbf{b}_j + \xi \mathbf{b}_{n+1} = 0 \qquad (1.12)$$

where $\mathbf{b}_k$ denotes the k-th row of the matrix (1.11). Since T is a circuit, there are constants $\zeta_k \neq 0$ so that $\mathbf{b}_1 = \sum_{k=2}^{q+1} \zeta_k \mathbf{b}_k$. Substituting this expression in (1.12) yields a dependence on the set $T_1 \cup J \cup \{n+1\}$, which is a contradiction.

Let $S = T \cup J \cup \{n + 1\}$. Then S_i is independent in type $\mathcal{T}$ for each i, $1 \leq i \leq q+1$. Since $\mathcal{T}'$ is a codimension-one degeneration of $\mathcal{T}$, if $T_i \in \mathrm{Dep}(\mathcal{T}')$, then $\mathsf{X}(\mathcal{T}')$ is locally defined by the vanishing of the submatrix Δ_{S_i} of (1.11) whose rows are specified by S_i in $\overline{\mathsf{X}}(\mathcal{T})$. If $T_j \in \mathrm{Dep}(\mathcal{T}')$ for $j \neq i$, then by Lemma 1.9.4, $T_k \in \mathrm{Dep}(\mathcal{T}')$ for every k. So, as above, $\mathsf{X}(\mathcal{T}')$ is locally defined by the vanishing of Δ_{S_k} in $\overline{\mathsf{X}}(\mathcal{T})$ for every k, $1 \leq k \leq q+1$. We will show that this is a contradiction by exhibiting a point in $\overline{\mathsf{X}}(\mathcal{T})$ for which Δ_{S_1} vanishes but Δ_{S_k} does not vanish for $k \neq 1$.

Assume that $J = (q+2, \dots, \ell+1)$. Then $\mathsf{X}(\mathcal{T})$ contains points of the form

$$\mathbf{b}(t) = \begin{pmatrix} 0 & I_q & 0 \\ 0 & v & 0 \\ 0 & 0 & I_{\ell-q} \\ F(t) & G(t) & H(t) \\ 1 & 0 & 0 \end{pmatrix}$$

where I_k is the $k \times k$ identity matrix, $v = (t\, 1 \cdots 1)$, and the submatrix $(F(t)\, G(t)\, H(t))$ is chosen so that $\mathrm{b}(t)$ satisfies the dependence and independence conditions of type $\mathcal{T}$ for each nonzero t. The point $\mathrm{b}(0)$ is in $\overline{X}(\mathcal{T})$, and for $1 \leq k \leq q+1$, $\Delta_{S_k}(\mathrm{b}(0))$ vanishes only for $k = 1$. Thus, if $T_i, T_j \in \mathrm{Dep}(\mathcal{T}')$, then there is a realizable type $\mathcal{T}''$ such that $\mathcal{T} > \mathcal{T}'' > \mathcal{T}'$. This contradicts the assumption that $\mathcal{T}$ covers $\mathcal{T}'$. $\qquad\square$

This result shows that $\mathrm{Dep}(\mathcal{T}', \mathcal{T})_q$ has no T-relevant element unless T has a codimension-one degeneration of Type II. In this case, there is a unique p so that T_p is the only T-relevant element in $\mathrm{Dep}(\mathcal{T}', \mathcal{T})_q$. It remains to consider T-relevant $S \in \mathrm{Dep}(\mathcal{T}', \mathcal{T})_{q+1}$.

Lemma 1.9.6. *Let $T \in \mathrm{Dep}(\mathcal{T})_{q+1}$ be a circuit. If all T-relevant $S \in \mathrm{Dep}(\mathcal{T}', \mathcal{T})_{q+1}$ belong to a family of a single type, then $m_S(\mathcal{T}') = 1$ for each such S.*

Proof. By relabeling the hyperplanes we may assume that $T = (U, n+1)$ where $U = (1, \ldots, q)$. Suppose the degeneration is of Type II so $(U, k) \in \mathrm{Dep}(\mathcal{T}', \mathcal{T})_{q+1}$ for some $k \in [n] - U$. Argue by contradiction. If $m_{(U,k)}(\mathcal{T}') = 2$, then in type $\mathcal{T}'$ there are two linearly independent vectors $\boldsymbol{\alpha} = (\alpha_1, \ldots, \alpha_q, \alpha_k)$ and $\boldsymbol{\beta} = (\beta_1, \ldots, \beta_q, \beta_k)$ which are annihilated by the rows of (1.11) specified by (U, k). If $\alpha_1 = 0$, then $(U_1, k) \in \mathrm{Dep}(\mathcal{T}')$. Since $(U, n+1)$ is a circuit in $\mathcal{T}$, and $(U_1, k) \notin \mathrm{Dep}(\mathcal{T})$ by Proposition 1.9.5, we have $(U_1, k) \in \mathrm{Dep}(\mathcal{T}', \mathcal{T})$ and hence $(U_1, k, n+1) \in \mathrm{Dep}(\mathcal{T}', \mathcal{T})$. This contradicts the assumption that all T-relevant sets S belong to a Type II family. If $\alpha_1 \neq 0$, then we use it to eliminate β_1 and find the same contradiction.

If the degeneration is of Type III, we may assume that $(U_1, p, n+1) \in \mathrm{Dep}(\mathcal{T}', \mathcal{T})$ with $p \in [n] - U$. Assuming that $m_{(U_1, p, n+1)}(\mathcal{T}') = 2$ leads to a similar argument. We consider the coefficient α_{n+1} and conclude that $(U_1, p) \in \mathrm{Dep}(\mathcal{T}', \mathcal{T})$ and hence $(U, p) \in \mathrm{Dep}(\mathcal{T}', \mathcal{T})$. This contradicts the assumption that all T-relevant sets S belong to a Type III family. $\qquad\square$

Lemma 1.9.7. *Let $T \in \mathrm{Dep}(\mathcal{T})_{q+1}$ be a circuit. Suppose T gives rise to codimension-one degenerations of both Type II and Type III. Then the Type II family is unique. For each Type III family there is a unique $p \in [n+1] - T$ so that (T_i, p) is also in the unique Type II family. We call (T_i, p) the intersection of these families. Moreover, $m_{(T_i, p)}(\mathcal{T}') = 2$ for each intersection and $m_S(\mathcal{T}') = 1$ for all other T-relevant S in these families.*

Proof. By relabeling the hyperplanes we may assume that $T = (U, n+1)$ where $U = (1, \ldots, q)$. If T gives rise to a Type II family, then it is of the form $\{(U, k) \mid k \in [n] - U\}$. By Proposition 1.9.5 there is a unique j for which $T_j \in \mathrm{Dep}(\mathcal{T}')_q$. We may assume that $j = q+1$ so that $T_{q+1} = U \in \mathrm{Dep}(\mathcal{T}')$. If T gives rise to two different families of Type II, then also some $(U_i, n+1) \in \mathrm{Dep}(\mathcal{T}')$, contradicting Proposition 1.9.5.

Suppose there is also a Type III family involving T. (There may be several Type III families involving T, but it will be clear from the proof that we may consider one Type III family at a time.) Let (U, p) be the intersection of the given Type II family and this Type III family. We show first that $m_{(U,p)}(T') = 2$. Since $U \in \mathrm{Dep}(T', T)$, it suffices to prove that row p is a linear combination of the rows specified by U in (1.11). Since $(U, n+1) \in \mathrm{Dep}(T)$, there is a vector $\boldsymbol{\alpha} = (\alpha_1, \ldots, \alpha_q, \alpha_{n+1}, 0)$ which is annihilated by the rows $(U, n+1, p)$ of (1.11). Since $(U, n+1)$ is a circuit, all $\alpha_i \neq 0$. This dependency holds also in type T'. Since $(U_1, p, n+1) \in \mathrm{Dep}(T')$, we also have a vector $\boldsymbol{\beta} = (0, \beta_2, \ldots, \beta_q, \beta_{n+1}, \beta_p)$ annihilated by the rows $(U, n+1, p)$ of (1.11) in type T'. We claim that $\beta_p \neq 0$, for otherwise we would have $(U_1, n+1) \in \mathrm{Dep}(T', T)$, contradicting Proposition 1.9.5. The vector $\beta_{n+1}\boldsymbol{\alpha} - \alpha_{n+1}\boldsymbol{\beta}$ provides the required dependence. Hence, $m_{(U,p)}(T') \geq 2$. Assuming that $m_{(U,p)}(T') > 2$ contradicts Proposition 1.9.5.

The fact that the other multiplicities in these families are 1 is established as in the proof of Lemma 1.9.6. $\qquad\square$

Theorem 1.9.8. *The endomorphism* $\tilde{\omega}(T', T)$ *satisfies* $\tilde{\omega}(T', T)(\mathsf{I}^\bullet(T)) \subset \mathsf{I}^\bullet(T)$ *if and only if for each* $K \subset [n]$ *and each circuit* $T \in \mathrm{Dep}(T)$ *we have* $\sum \tilde{\omega}_S(a_K r_T) \in \mathsf{I}^\bullet(T)$ *where the sum is over* T-*relevant* S *in a single type of codimension one degeneration involving* T.

Proof. This is clear if T is involved in a single type. If more types appear, then Lemma 1.9.7 shows that each S is the intersection of at most two types. Furthermore, all such intersections have multiplicity 2, so the corresponding S may be considered individually in their respective types, each time with multiplicity 1. $\qquad\square$

1.10 Ideal Invariance

In this section we show that a suitable sum of universal endomorphisms $\tilde{\omega}_S$ in the Aomoto complex of the general position type induces an endomorphism in the Aomoto complex of of type T for all pairs of types T, T' where T covers T'.

Theorem 1.10.1. *If* T *covers* T', *then* $\tilde{\omega}(T', T)(\mathsf{I}^\bullet(T)) \subset \mathsf{I}^\bullet(T)$.

Proof. It follows from Theorem 1.9.8 that we may argue on the different types independently. It suffices to show that for every circuit $T \in \mathrm{Dep}(T)$, every k-tuple K, and every degeneration T' of T, we have $\tilde{\omega}(T')(a_K r_T) \in \mathsf{I}^\bullet(T)$. As before, we must consider several cases. Note that $n+1 \notin K$, and we agree to use the same symbol for the underlying set. Similarly, if L is a set which does not contain $n+1$, then we write L for the corresponding tuple in the standard order.

The following identity will be useful in several parts. If $J \subset [n]$, then

$$\left(\sum_{m \in J} y_m a_m \right) \partial a_J = \left(\sum_{m \in J} y_m \right) a_J.$$

Case 1: $T \in \mathrm{Dep}(\mathcal{T})_{q+1}$ is a circuit with $n + 1 \in T$

Write $T = (U, n + 1)$, and assume that $U = (n - q + 1, \ldots, n)$. First assume $T \subset S \in \mathrm{Dep}(\mathcal{T}')$. Clearly, $\{K, T\} \in \mathrm{Dep}(\mathcal{T}')$. Let $L = [n] \setminus \{K \cup U\}$. We get

$$\tilde{\omega}^{k+q}_{\{K,U,n+1\}}(a_K a_U) = -\Big(\sum_{j \in L} y_j\Big) a_K a_U.$$

For every $j \in K$, $\{K_j, T\} \in \mathrm{Dep}(\mathcal{T}')$. Here

$$\tilde{\omega}^{k+q}_{\{K_j,T\}}(a_K a_U) = (-1)^j a_\mathbf{y} a_{K_j} a_U.$$

Similarly, for every $j \in K$ and every $m \in L$, $\{K_j, m, T\} \in \mathrm{Dep}(\mathcal{T}')$. Here

$$\tilde{\omega}^{k+q}_{\{K_j,m,U,n+1\}}(a_K a_U) = a_m a_{K_j} a_U.$$

In the remaining parts of this case we may assume that $T \not\subset S$ for $S \in \mathrm{Dep}(\mathcal{T}')$.

Case 1.1

If $|S \cap \{K, U\}| < k + q - 1$ for all other $S \in \mathrm{Dep}(\mathcal{T}')$, then we are done by Lemma 1.9.2. If there exists $S \in \mathrm{Dep}(\mathcal{T}')$ with $|S \cap \{K, U\}| \geq k + q - 1$ and $T \not\subset S$, then $S = \{K, T_p, m\}$ with $m \in [n + 1] \setminus T$. The classification implies that (T_p, m) is in Type II or III, and all the other members of that type must also be in $\mathrm{Dep}(\mathcal{T}')$.

Case 1.2

Suppose (T_p, m) belongs to Type II. Then p is fixed. If $p \neq q + 1$, then we may assume that $p = 1$. Thus $\mathrm{Dep}(\mathcal{T}')$ contains $S_m = \{m, K, T_1\}$ for all $m \in L$. Since every S_m contains $F = \{K, T_1\}$, we conclude that $F \in \mathrm{Dep}(\mathcal{T}')$. Here $\tilde{\omega}^{k+q}_F(a_K a_U) = (-1)^{k+1} a_\mathbf{y} a_K a_{U_1}$ and $\tilde{\omega}^{k+q}_{S_m}(a_K a_U) = (-1)^k y_m a_m a_K a_{U_1}$. Thus

$$\Big(\tilde{\omega}^{k+q}_F + \sum_{m \in L} \tilde{\omega}^{k+q}_{S_m}\Big)(a_K a_U) = -y_{n-q+1} a_K a_U.$$

If $p = q + 1$, then $\mathrm{Dep}(\mathcal{T}')$ contains $S_m = \{m, K, U\}$ for all $m \in L$. Since every S_m contains $F = \{K, U\}$, we conclude that $F \in \mathrm{Dep}(\mathcal{T}')$. Here $\tilde{\omega}^{k+q}_F(a_K a_U) = a_\mathbf{y} \partial(a_K a_U)$ and $\tilde{\omega}^{k+q}_{S_m}(a_K a_U) = y_m \partial a_{(m,K,U)} = y_m a_K a_U - y_m a_m \partial(a_K a_U)$. Thus

$$\Big(\tilde{\omega}^{k+q}_F + \sum_{m \in L} \tilde{\omega}^{k+q}_{S_m}\Big)(a_K a_U) = \Big(\sum_{j \in [n]} y_j\Big) a_K a_U.$$

Case 1.3

If (T_p, m) belongs to Type III, then for some fixed $m \in L$, $\mathrm{Dep}(\mathcal{T}')$ contains the sets $S_p = \{m, K, T_p\}$ for all p with $1 \le p \le q+1$. If $p \ne q+1$, then $\tilde{\omega}_{S_p}^{k+q}(a_K a_U) = (-1)^{k+p+1} y_m a_{(m,K,U)_{k+p+1}} = (-1)^{k+p+1} y_m a_m a_K a_{U_p}$. Thus $\sum_{p=1}^{q} \tilde{\omega}_{S_p}^{k+q}(a_K a_U) = (-1)^k y_m a_m a_K \partial a_U$. If $p = q+1$, then $\tilde{\omega}_{\{m,K,U\}}^{k+q}(a_K a_U) = y_m \partial a_{(m,K,U)} = y_m a_K a_U - y_m a_m(\partial a_K) a_U + (-1)^{k+1} y_m a_m a_K \partial a_U$. Thus

$$\sum_{p=1}^{q+1} \tilde{\omega}_{S_p}^{k+q}(a_K a_U) = y_m(a_K - a_m \partial a_K) a_U.$$

This completes the argument in **Case 1**.

Case 2: $T \in \mathrm{Dep}(\mathcal{T})_{q+1}$ is a circuit with $n + 1 \notin T$

Here we may assume that $T = (1, \ldots, q+1)$. We note that $a_T = a_1(\partial a_T) \in I(\mathcal{T})$ and hence $\partial(a_K a_T) \in I(\mathcal{T})$. First assume $T \subset S \in \mathrm{Dep}(\mathcal{T}')$. Clearly, $\{K, T\} \in \mathrm{Dep}(\mathcal{T}')$ and we have

$$\tilde{\omega}_{\{K,T\}}^{k+q}(a_K \partial a_T) = \sum_{j \in T}(-1)^{j-1}\tilde{\omega}_{\{K,T\}}^{k+q}(a_K a_{T_j}) = \sum_{j \in T}(-1)^{j-1} y_j \partial a_{(j,K,T_j)}$$
$$= (-1)^k(\sum_{j \in T} y_j)\partial(a_K a_T).$$

For $j \in K$, $\{K_j, T\} \in \mathrm{Dep}(\mathcal{T}')$, but $\tilde{\omega}_{\{K_j,T\}}^{k+q}(a_K \partial a_T) = 0$. Let $L = [n] \setminus \{K \cup T\}$. For every $j \in K$ and every $m \in L$, $\{K_j, m, T\} \in \mathrm{Dep}(\mathcal{T}')$, but only $m = n+1$ gives a nonzero term:

$$\tilde{\omega}_{\{K_j,T,n+1\}}^{k+q}(a_K \partial a_T) = (-1)^k(\sum_{s \in T} y_s)a_{K_j} a_T.$$

In the remaining parts of this case, we may assume that $T \not\subset S$ for $S \in \mathrm{Dep}(\mathcal{T}')$.

Case 2.1

If $|S \cap \{K, T_j\}| < k+q-1$ for all other $S \in \mathrm{Dep}(\mathcal{T}')$ for all j, then we are done by Lemma 1.9.2. If there is a Type I degeneration S so that $\tilde{\omega}_S$ is not zero on some term of ∂a_T, then $|S \cap \{K, T_j\}| = k+q-1$, and it follows from the definition of formal connections that $n+1 \in S$. In this case we may assume that $S = \{K, 3, 4, \ldots, q+1, m, n+1\}$ where $m \in [n] \setminus \{K, T\}$. Lemma 1.9.3 shows that $\tilde{\omega}_S(\partial a_T) = 0$.

Case 2.2

Suppose $(T_p, m) \in \text{Dep}(\mathcal{T}')$ with $m \in [n+1] \setminus T$ belongs to Type II. Then p is fixed, and we may assume that $p = q+1$. Thus $S_m = \{m, K, T_{q+1}\} \in \text{Dep}(\mathcal{T}')$ for all $m \notin T$. Since every S_m contains $F = \{K, T_{q+1}\}$, we conclude that $F \in \text{Dep}(\mathcal{T}')$. We have

$$\tilde{\omega}_F^{k+q}(a_K \partial a_T) = (-1)^q a_{\mathbf{y}} \partial(a_K a_{T_{q+1}}).$$

Since $T_{q+1} \in \text{Dep}(\mathcal{T}')$, we also have $\{K_j, T_{q+1}, n+1\} \in \text{Dep}(\mathcal{T}')$ for all $j \in K$. Here $\tilde{\omega}_{\{K_j, T_{q+1}, n+1\}}^{k+q}(a_K \partial a_T) = (-1)^{q+j} a_{\mathbf{y}} a_{K_j} a_{T_{q+1}}$. Thus

$$\sum_{j \in K} \tilde{\omega}_{\{K_j, T_{q+1}, n+1\}}^{k+q}(a_K \partial a_T) = (-1)^{q+1} a_{\mathbf{y}}(\partial a_K) a_{T_{q+1}}.$$

If $m \neq n+1$, then $\tilde{\omega}_{\{m, F\}}^{k+q}(a_K \partial a_T) = (-1)^q y_m \partial(a_{m,K,T_{q+1}})$. Note that $m \notin T$ by the classification, and $m \notin K$ follows from the expression. Thus $m \in L$ and we get

$$\sum_{m \in L} \tilde{\omega}_{\{m, F\}}^{k+q}(a_K \partial a_T) = (-1)^q \sum_{m \in L} y_m \partial(a_{m,K,T_{q+1}}).$$

Let $m = n+1$. We have $\tilde{\omega}_{\{F, n+1\}}^{k+q}(a_K \partial a_T) = \sum_{p=1}^{q+1}(-1)^{p-1} \tilde{\omega}_{\{F, n+1\}}^{k+q}(a_K a_{T_p})$. For $p \neq q+1$, $\tilde{\omega}_{\{F, n+1\}}^{k+q}(a_K a_{T_p}) = (-1)^{p+q} y_p a_K a_{T_{q+1}}$ and $\tilde{\omega}_{\{F, n+1\}}^{k+q}(a_K a_{T_{q+1}}) = -(\sum_{p \notin K \cup T_{q+1}} y_p) a_K a_{T_{q+1}}$. Thus

$$\tilde{\omega}_{\{F, n+1\}}^{k+q}(a_K \partial a_T) = (-1)^{q+1}\Big(\sum_{j \in [n] \setminus K} y_j \Big) a_K a_{T_{q+1}}.$$

Similarly, for every $j \in K$ and $s \in L$, $\{K_j, s, T_{q+1}, n+1\} \in \text{Dep}(\mathcal{T}')$. Here $\tilde{\omega}_{\{K_j, s, T_{q+1}, n+1\}}^{k+q}(a_K a_{T_p}) = (-1)^{q+j+1} y_s a_s a_{K_j} a_{T_{q+1}}$. Thus

$$\sum_{s \in L} \sum_{j \in K} \tilde{\omega}_{\{K_j, s, T_{q+1}, n+1\}}^{k+q}(a_K a_{T_p}) = (-1)^q \Big(\sum_{s \in L} y_s a_s \Big)(\partial a_K) a_{T_{q+1}}.$$

Summing over all dependent sets in Type II, we must compute $\xi(a_K \partial a_T)$, where $\xi =$

$$\tilde{\omega}_F^{k+q} + \sum_{j \in K} \tilde{\omega}_{\{K_j, T_{q+1}, n+1\}}^{k+q} + \sum_{m \in L} \tilde{\omega}_{\{m, F\}}^{k+q} + \tilde{\omega}_{\{F, n+1\}}^{k+q} + \sum_{s \in L} \sum_{j \in K} \tilde{\omega}_{\{K_j, s, T_{q+1}, n+1\}}^{k+q}.$$

We get

$$\xi(a_K \partial a_T) = (-1)^q a_{\mathbf{y}} \partial(a_K a_{T_{q+1}}) + (-1)^{q+1} a_{\mathbf{y}}(\partial a_K) a_{T_{q+1}}$$
$$+ (-1)^q \sum_{m \in L} y_m \partial(a_{m,K,T_{q+1}}) + (-1)^{q+1} \Big(\sum_{j \in [n] \setminus K} y_j \Big) a_K a_{T_{q+1}}$$
$$+ (-1)^q \Big(\sum_{s \in L} y_s a_s \Big) (\partial a_K) a_{T_{q+1}}$$

$$= (-1)^q \Big(\sum_{j \in K \cup T_{q+1}} y_j + \sum_{m \in L} y_m - \sum_{j \in [n] \setminus K} y_j \Big) a_K a_{T_{q+1}}$$
$$+ (-1)^q \Big(y_{q+1} a_{q+1} + \sum_{m \in L} y_m a_m - \sum_{m \in L} y_m a_m \Big) \partial(a_K a_{T_{q+1}})$$
$$+ (-1)^q \Big(- \sum_{j \in [n] \setminus T_{q+1}} y_j a_j + \sum_{m \in L} y_m a_m \Big) (\partial a_K) a_{T_{q+1}}$$

$$= (-1)^{q+1} y_{q+1} \big[a_K a_{T_{q+1}} - a_{q+1} \partial(a_K a_{T_{q+1}}) + a_{q+1} (\partial a_K) a_{T_{q+1}} \big]$$
$$= -y_{q+1} a_K \partial a_T.$$

Case 2.3

If (T_p, m) belongs to Type III, then for some fixed $m \notin T$, $\mathrm{Dep}(T')$ contains the sets $S_p = \{m, K, T_p\}$ for all $p \in T$. If $m \neq n+1$, then

$$\sum_{p \in T} \tilde{\omega}_{S_p}^{k+q}(a_K \partial a_T) = \sum_{p \in T} (-1)^{p-1} y_m \partial(a_m a_K a_{T_p}) = y_m (a_K - a_m \partial a_K) \partial a_T.$$

For $m = n+1$ we need the formulas

$$\tilde{\omega}_{\{K, T_p, n+1\}}^{k+q}(a_K a_{T_s}) = \begin{cases} (-1)^{p+s-1} y_s a_K a_{T_p} & \text{if } s \neq p, \\ -\Big(\sum_{j \in [n] \setminus \{K, T_p\}} y_j \Big) a_K a_{T_p} & \text{if } s = p. \end{cases}$$

We get $\tilde{\omega}_{\{K, T_p, n+1\}}^{k+q}(a_K \partial a_T) = (-1)^p \big(\sum_{j \in [n] \setminus K} y_j \big) a_K a_{T_p}$. Thus

$$\sum_{p \in T} \tilde{\omega}_{\{K, T_p, n+1\}}^{k+q}(a_K \partial a_T) = -\Big(\sum_{j \in [n] \setminus K} y_j \Big) a_K \partial a_T.$$

This completes the argument in **Case 2**, and hence the proof of Theorem 1.10.1. □

Combinatorial Gauss-Manin Connections

Theorem 1.10.1 implies that there is a commutative diagram

$$\begin{array}{ccccc}
(\mathsf{I}^\bullet(T), a_{\mathbf{y}}) & \xrightarrow{\ \iota\ } & (\mathsf{A}^\bullet(\mathcal{G}), a_{\mathbf{y}}) & \xrightarrow{\ p\ } & (\mathsf{A}^\bullet(T), a_{\mathbf{y}}) \\
\Big\downarrow {\scriptstyle \tilde{\omega}(T',T)|_{\mathsf{I}^\bullet(T)}} & & \Big\downarrow {\scriptstyle \tilde{\omega}(T',T)} & & \Big\downarrow {\scriptstyle \omega(T',T)} \\
(\mathsf{I}^\bullet(T), a_{\mathbf{y}}) & \xrightarrow{\ \iota\ } & (\mathsf{A}^\bullet(\mathcal{G}), a_{\mathbf{y}}) & \xrightarrow{\ p\ } & (\mathsf{A}^\bullet(T), a_{\mathbf{y}})
\end{array}$$

where $\iota : \mathsf{I}^\bullet(\mathcal{T}) \to \mathsf{A}^\bullet(\mathcal{G})$ is the inclusion, $p : \mathsf{A}^\bullet(\mathcal{G}) \to \mathsf{A}^\bullet(\mathcal{T})$ is the projection provided by the respective **nbc** bases, and $\omega(\mathcal{T}', \mathcal{T})$ is the induced map. It follows that for given weights $\boldsymbol{\lambda}$, the specialization $\mathbf{y} \mapsto \boldsymbol{\lambda}$ in the chain endomorphism $\omega(\mathcal{T}', \mathcal{T})$ defines chain endomorphisms $\omega_{\boldsymbol{\lambda}}^q(\mathcal{T}', \mathcal{T}) : A^q(\mathcal{T}) \to A^q(\mathcal{T})$ for $0 \le q \le \ell$. Let $\rho^q : A^q \twoheadrightarrow H^q(A^\bullet(\mathcal{T}), a_{\boldsymbol{\lambda}})$ be the natural projection. It follows that $\omega_{\boldsymbol{\lambda}}^q(\mathcal{T}', \mathcal{T})$ induces an endomorphism

$$\Omega_{\mathcal{C}}^q(\mathcal{T}', \mathcal{T}) : H^q(A^\bullet(\mathcal{T}), a_{\boldsymbol{\lambda}}) \to H^q(A^\bullet(\mathcal{T}), a_{\boldsymbol{\lambda}})$$

determined by the equation $\rho^q \circ \omega_{\boldsymbol{\lambda}}^q(\mathcal{T}', \mathcal{T}) = \Omega_{\mathcal{C}}^q(\mathcal{T}', \mathcal{T}) \circ \rho^q$. We call $\Omega_{\mathcal{C}}^q(\mathcal{T}', \mathcal{T})$ a *combinatorial Gauss-Manin connection* endomorphism in Orlik-Solomon algebra cohomology. The terminology will be explained in Section 2.5.

1.11 Examples

The Selberg Arrangement

We illustrate the definitions and results of Chapter 1 in the Selberg arrangement. Figure 1.4 shows the arrangement $\mathcal{A}$ and its partially ordered set $L(\mathcal{A})$.

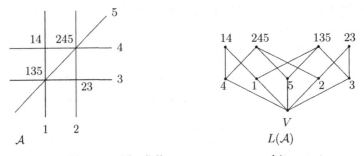

Fig. 1.4. The Selberg arrangement and its poset

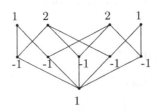

Fig. 1.5. Values of the Möbius function

The values of the Möbius function are given in Figure 1.5. Thus $\pi(\mathcal{A}, t) = 1 + 5t + 6t^2$ and there are 12 chambers in the complement. Furthermore $\beta(\mathcal{A}) = (-1)^2(1 - 5 + 6) = 2$ is the number of bounded chambers.

Next we construct the Orlik-Solomon algebra $A(\mathcal{A})$. Note that the subsets (12) and (34) have $\cap S = \emptyset$ and the circuits are (135) and (245). Thus the algebra relations are

$$a_{12} = a_{34} = 0, \quad a_{35} - a_{15} + a_{13} = 0, \quad a_{45} - a_{25} + a_{24} = 0.$$

It follows that the broken circuits are (35) and (45). Thus $A(\mathcal{A})$ has the following **nbc** basis:

$$1 \qquad a_1, a_2, a_3, a_4, a_5 \qquad a_{13}, a_{14}, a_{15}, a_{23}, a_{24}, a_{25}$$

The complex $\mathrm{NBC}(\mathcal{A})$ is shown in Figure 1.6. Its 1-simplexes are the **nbc** frames.

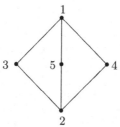

Fig. 1.6. The NBC complex of the Selberg arrangement

Here $\beta\mathbf{nbc} = \{(24), (25)\}$ and the cohomology classes $[(24)^*]$ and $[(25)^*]$ form a basis for $H^1(\mathrm{NBC}(\mathcal{A}))$.

Next consider the complex $(A(\mathcal{A}), a_\lambda)$. If at least one $\lambda_i \neq 0$, then the map $a_\lambda : A^0 \to A^1$ is a monomorphism. The table below describes $a_\lambda : A^1 \to A^2$ in terms of these **nbc** generators. We use the notation $\lambda_J = \sum_{j \in J} \lambda_j$.

	a_{13}	a_{14}	a_{15}	a_{23}	a_{24}	a_{25}
a_1	$-\lambda_3$	$-\lambda_4$	$-\lambda_5$			
a_2				$-\lambda_3$	$-\lambda_4$	$-\lambda_5$
a_3	λ_{15}		$-\lambda_5$	λ_2		
a_4		λ_1			λ_{25}	$-\lambda_5$
a_5	$-\lambda_3$		λ_{13}		$-\lambda_4$	λ_{24}

The remaining calculations in this example were provided by Dan Cohen. Let $\mathcal{G}$ be the combinatorial type of a general position arrangement of five lines in $\mathbb{C}^2$. The **nbc** bases for the Orlik-Solomon algebras $A^\bullet(\mathcal{G})$ and $A^\bullet(\mathcal{T})$ give rise to bases for the corresponding Aomoto complexes. The Aomoto complex $(A^\bullet(\mathcal{G}), a_\mathbf{y})$ is given by

$$A^0(\mathcal{G}) \xrightarrow{a_y} A^1(\mathcal{G}) \xrightarrow{a_y} A^2(\mathcal{G})$$

where $A^0(\mathcal{G}) = R$, $A^1(\mathcal{G}) = R^5$, $A^2(\mathcal{G}) = R^{10}$, and the matrices of the boundary maps are $\begin{bmatrix} y_1 & y_2 & y_3 & y_4 & y_5 \end{bmatrix}$ and

$$\begin{bmatrix} -y_2 & -y_3 & -y_4 & -y_5 & 0 & 0 & 0 & 0 & 0 & 0 \\ y_1 & 0 & 0 & 0 & -y_3 & -y_4 & -y_5 & 0 & 0 & 0 \\ 0 & y_1 & 0 & 0 & y_2 & 0 & 0 & -y_4 & -y_5 & 0 \\ 0 & 0 & y_1 & 0 & 0 & y_2 & 0 & y_3 & 0 & -y_5 \\ 0 & 0 & 0 & y_1 & 0 & 0 & y_2 & 0 & y_3 & y_4 \end{bmatrix}.$$

The Aomoto complex $A^\bullet(\mathcal{T})$ of the Selberg arrangement is given by

$$A^0(\mathcal{T}) \xrightarrow{a_y} A^1(\mathcal{T}) \xrightarrow{a_y} A^2(\mathcal{T})$$

where $A^0(\mathcal{T}) = R$, $A^1(\mathcal{T}) = R^5$, and $A^2(\mathcal{T}) = R^6$. The matrix of the boundary map $A^1(\mathcal{T}) \xrightarrow{a_y} A^2(\mathcal{T})$ is

$$\begin{bmatrix} -y_3 & -y_4 & -y_5 & 0 & 0 & 0 \\ 0 & 0 & 0 & -y_3 & -y_4 & -y_5 \\ y_{15} & 0 & -y_5 & y_2 & 0 & 0 \\ 0 & y_1 & 0 & 0 & y_{25} & -y_5 \\ -y_3 & 0 & y_{13} & 0 & -y_4 & y_{24} \end{bmatrix}$$

where $y_J = \sum_{j \in J} y_j$.

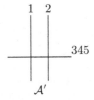

Fig. 1.7. The Selberg arrangement and a degeneration

In Figure 1.7 the multi-arrangement $\mathcal{A}'$ is a codimension-one degeneration of type $\mathcal{T}'$ of the Selberg arrangement of type $\mathcal{T}$. Here $\text{Dep}(\mathcal{T}) = \{135, 245, 126, 346\}$ and

$$\text{Dep}(\mathcal{T}', \mathcal{T}) = \{34, 35, 45, 134, 145, 234, 235, 345, 356, 456\}.$$

Since lines 3, 4, and 5 coincide in type $\mathcal{T}'$, it follows from the definition that $m_{345} = 2$. Note that $T = (346) \in \text{Dep}(\mathcal{T})$ gives rise to the Type II family $\{341, 342, 345\}$. Here $p = 3$ is the position of the entry deleted in T. It is replaced one at a time by the elements not in T. Note also that $T = (346) \in \text{Dep}(\mathcal{T})$ gives rise to the Type III family $\{465, 365, 345\}$. Here $m = 5$, so we insert 5 and delete one at a time each entry in T. The intersection of these two families is $(T_3, 5) = (345)$. It follows from the definition that $\tilde{\omega}(T', T) : A^{\bullet}(\mathcal{G}) \to A^{\bullet}(\mathcal{G})$ is given by

$$\tilde{\omega}(T', T) = \tilde{\omega}_{34} + \tilde{\omega}_{35} + \tilde{\omega}_{45} + \tilde{\omega}_{134} + \tilde{\omega}_{145} + \tilde{\omega}_{234} + \tilde{\omega}_{235} + 2\tilde{\omega}_{345} + \tilde{\omega}_{356} + \tilde{\omega}_{456}.$$

Note that the multiplicity $m_{345} = 2$ enters here. The matrices of this chain endomorphism are $\tilde{\omega}^0(T', T) = 0$,

$$\tilde{\omega}^1(T', T) = \begin{bmatrix} 0 & 0 & 0 & 0 & 0 \\ 0 & 0 & 0 & 0 & 0 \\ 0 & 0 & y_{45} & -y_5 & -y_5 \\ 0 & 0 & -y_3 & y_{35} & -y_5 \\ 0 & 0 & -y_3 & -y_4 & y_{34} \end{bmatrix},$$

$$\tilde{\omega}^2(T', T) = \begin{bmatrix} 0 & 0 & 0 & 0 & 0 & 0 & 0 & 0 & 0 & 0 \\ 0 & y_{45} & -y_4 & -y_5 & 0 & 0 & 0 & y_4 & 0 & 0 \\ 0 & -y_3 & y_{35} & -y_5 & 0 & 0 & 0 & -y_3 & 0 & 0 \\ 0 & -y_3 & -y_4 & y_{34} & 0 & 0 & 0 & 0 & 0 & 0 \\ 0 & 0 & 0 & 0 & y_{45} & -y_4 & -y_5 & y_4 & 0 & 0 \\ 0 & 0 & 0 & 0 & -y_3 & y_3 & 0 & -y_3 & 0 & -y_5 \\ 0 & 0 & 0 & 0 & -y_3 & 0 & y_3 & 0 & 0 & y_4 \\ 0 & 0 & 0 & 0 & 0 & 0 & 0 & y_{[5]} + y_5 & 0 & 0 \\ 0 & 0 & 0 & 0 & 0 & 0 & 0 & -y_4 & y_{345} & 0 \\ 0 & 0 & 0 & 0 & 0 & -y_2 & y_2 & y_3 & 0 & y_{345} - y_2 \end{bmatrix}.$$

The projection $p : A^{\bullet}(\mathcal{G}) \to A^{\bullet}(T)$ is given in terms of the respective **nbc** bases by

$$p(a_J) = \begin{cases} 0 & \text{if } J = (12) \text{ or } J = (34), \\ a_{15} - a_{13} & \text{if } J = (35), \\ a_{25} - a_{24} & \text{if } J = (45), \\ a_J & \text{otherwise.} \end{cases}$$

This yields the induced endomorphism $\omega(T', T) : A^{\bullet}(T) \to A^{\bullet}(T)$, given explicitly by $\omega^0(T', T) = 0$, $\omega^1(T', T) = \tilde{\omega}^1(T', T)$, and

$$\omega^2(T', T) = \begin{bmatrix} y_{45} & -y_4 & -y_5 & 0 & 0 & 0 \\ -y_3 & y_{35} & -y_5 & 0 & 0 & 0 \\ -y_3 & -y_4 & y_{34} & 0 & 0 & 0 \\ 0 & 0 & 0 & y_{45} & -y_4 & -y_5 \\ 0 & 0 & 0 & -y_3 & y_{35} & -y_5 \\ 0 & 0 & 0 & -y_3 & -y_4 & y_{34} \end{bmatrix}.$$

The dense edges of $\mathcal{A}$ are $\{1, 2, 3, 4, 5, 135, 245\}$. The additional dense edges in its projective closure $\mathcal{A}_{\infty}$ with infinite hyperplane H_6 are $\{6, 126, 346\}$. Recall that $\lambda_6 = -\lambda_{12345}$. The weights satisfy the conditions of Theorem 1.6.7 provided the following are not zero:

$$\lambda_1, \lambda_2, \lambda_3, \lambda_4, \lambda_5, \lambda_6, \lambda_{126}, \lambda_{135}, \lambda_{245}, \lambda_{346}.$$

In this case

$$\zeta(\{24\}) = (\lambda_2 a_2 + \lambda_4 a_4 + \lambda_5 a_5)\lambda_4 a_4 = \lambda_2\lambda_4 a_{24} - \lambda_4\lambda_5 a_{45},$$
$$\zeta(\{25\}) = (\lambda_2 a_2 + \lambda_4 a_4 + \lambda_5 a_5)\lambda_5 a_5 = \lambda_2\lambda_5 a_{25} + \lambda_4\lambda_5 a_{45}.$$

Using the Orlik-Solomon relation $a_{45} = a_{25} - a_{24}$ shows that $\{\eta_{24} = \lambda_2\lambda_4 a_{24}, \eta_{25} = \lambda_2\lambda_5 a_{25}\}$ is a basis for the only nonvanishing group $H^2(A, a_\lambda)$. The projection $\rho^2 : A^2(\mathcal{T}) \twoheadrightarrow H^2(A^\bullet(\mathcal{T}), a_\lambda)$ is given by

$$\rho^2(a_{ij}) = \begin{cases} \dfrac{(\lambda_1\lambda_2 + \lambda_2\lambda_3 + \lambda_3\lambda_5)\eta_{24} + (\lambda_1\lambda_2 - \lambda_3\lambda_4)\eta_{25}}{\lambda_1\lambda_2\lambda_3\lambda_{135}} & \text{if } (ij) = (13), \\[2ex] \dfrac{-\lambda_{25}\eta_{24} + \lambda_4\eta_{25}}{\lambda_1\lambda_2\lambda_4} & \text{if } (ij) = (14), \\[2ex] \dfrac{\lambda_5\lambda_{125}\eta_{24} - (\lambda_1\lambda_2 + \lambda_1\lambda_4 + \lambda_4\lambda_5)\eta_{25}}{\lambda_1\lambda_2\lambda_5\lambda_{135}} & \text{if } (ij) = (15), \\[2ex] -\dfrac{\eta_{24} + \eta_{25}}{\lambda_2\lambda_3} & \text{if } (ij) = (23), \\[2ex] \dfrac{\eta_{24}}{\lambda_2\lambda_4} & \text{if } (ij) = (24), \\[2ex] \dfrac{\eta_{25}}{\lambda_2\lambda_5} & \text{if } (ij) = (25). \end{cases}$$

A calculation with the endomorphism $\omega_\lambda^2(\mathcal{T}', \mathcal{T}) = \omega^2(\mathcal{T}', \mathcal{T})|_{y \mapsto \lambda}$ and the projection ρ^2 yields the combinatorial Gauss-Manin connection

$$\Omega_{\mathcal{C}}^2(\mathcal{T}', \mathcal{T}) = \begin{bmatrix} \lambda_{345} & 0 \\ 0 & \lambda_{345} \end{bmatrix}.$$

Next consider λ which give rise to cohomology in other dimensions. The kernel of the map $a_\lambda : A^1 \to A^2$ has dimension two if $\lambda_1 = \lambda_4$, $\lambda_2 = \lambda_3$, $\lambda_5 = \lambda_6$, and $\lambda_{1,2,5} = 0$. If λ is nontrivial, then $\lambda_1 \neq 0$ or $\lambda_2 \neq 0$. For such weights, one can check that $a_1 - a_2 - a_3 + a_4$ generates $H^1(A^\bullet(\mathcal{T}), a_\lambda)$ and that the combinatorial Gauss-Manin connection $\Omega_{\mathcal{C}}^1(\mathcal{T}', \mathcal{T}) : H^1(A^\bullet(\mathcal{T}), a_\lambda) \to H^1(A^\bullet(\mathcal{T}), a_\lambda)$ is trivial. One can also show that, for an appropriate choice of basis for $H^2(A^\bullet(\mathcal{T}), a_\lambda)$, the projection $A^2(\mathcal{T}) \twoheadrightarrow H^2(A^\bullet(\mathcal{T}), a_\lambda)$ has matrix

$$\begin{bmatrix} \lambda_1 & -\lambda_1 & \lambda_2 \\ -\lambda_2 & \lambda_2 & \lambda_2 \\ 0 & 0 & \lambda_2 \\ \lambda_1 & -\lambda_1 & -\lambda_1 \\ \lambda_1 & \lambda_2 & \lambda_2 \\ \lambda_1 & 0 & 0 \end{bmatrix}$$

and that the combinatorial Gauss-Manin connection $\Omega_{\mathcal{C}}^2(\mathcal{T}', \mathcal{T})$ induced by $\omega_\lambda(\mathcal{T}', \mathcal{T})$ is also trivial.

Four Lines

Let T be the combinatorial type of the arrangement $\mathcal{A}$ of four lines in $\mathbb{C}^2$ depicted in Figure 1.8. Here $\mathsf{X}(T)$ has codimension one in $(\mathbb{CP}^2)^4 = \overline{\mathsf{X}}(\mathcal{G})$. The combinatorial types T_i of the (multi-)arrangements $\mathcal{A}_i$ shown in Figure 1.8 are codimension-one degenerations of T corresponding to irreducible components of the divisor $\mathsf{D}(T) = \overline{\mathsf{X}}(T) \setminus \mathsf{X}(T)$.

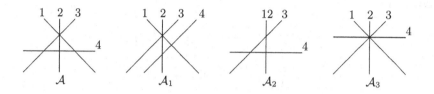

Fig. 1.8. Four lines and three degenerations

For these degenerations we have

$$\mathrm{Dep}(T_1, T) = \{345\},$$
$$\mathrm{Dep}(T_2, T) = \{12, 124, 125\},$$
$$\mathrm{Dep}(T_3, T) = \{124, 134, 234\}.$$

Here T_1 is a degeneration of Type I, T_2 is a degeneration of Type II, with $p = 3$, and T_3 is a degeneration of Type III, with $m = 4$. The corresponding endomorphisms $\tilde{\omega}(T_i, T)$ of the Aomoto complex $\mathsf{A}(\mathcal{G})$ of a general position arrangement of four lines may be calculated as follows. By definition

$$\tilde{\omega}(T_1, T) = \tilde{\omega}_{345}, \quad \tilde{\omega}(T_2, T) = \tilde{\omega}_{12} + \tilde{\omega}_{124} + \tilde{\omega}_{125}, \quad \tilde{\omega}(T_3, T) = \tilde{\omega}_{124} + \tilde{\omega}_{134} + \tilde{\omega}_{234},$$

and the individual $\tilde{\omega}_S$ are defined in Section 1.8. Let $p : \mathsf{A}^\bullet(\mathcal{G}) \to \mathsf{A}^\bullet(T)$ be the natural projection given in the **nbc** bases by

$$p(a_J) = \begin{cases} a_{13} - a_{12} & \text{if } J = (23), \\ a_J & \text{otherwise.} \end{cases}$$

By Theorem 1.10.1 the chain endomorphisms $\tilde{\omega}(T_j, T)$ of $\mathsf{A}^\bullet(\mathcal{G})$ induce chain endomorphisms $\omega(T_j, T)$ of $\mathsf{A}^\bullet(T)$. A calculation with the projection p reveals that these are given by $\omega^1(T_j, T) = \tilde{\omega}^1(T_j, T)$ for each j, and

$$\omega^2(T_1, T) = \begin{bmatrix} 0 & 0 & 0 & 0 & 0 \\ 0 & 0 & 0 & 0 & -y_4 \\ 0 & 0 & 0 & 0 & y_3 \\ 0 & 0 & 0 & 0 & y_3 \\ 0 & 0 & 0 & 0 & -y_{12} \end{bmatrix}, \quad \omega^2(T_2, T) = \begin{bmatrix} y_{12} & 0 & 0 & 0 & 0 \\ y_2 & 0 & 0 & 0 & 0 \\ 0 & 0 & y_2 & -y_2 & 0 \\ 0 & 0 & -y_1 & y_1 & 0 \\ 0 & 0 & 0 & 0 & 0 \end{bmatrix},$$

$$\omega^2(\mathcal{T}_3, \mathcal{T}) = \begin{bmatrix} y_4 & 0 & -y_4 & y_4 & 0 \\ 0 & y_4 & -y_4 & 0 & y_4 \\ -y_2 & -y_3 & y_{23} & -y_2 & -y_3 \\ y_{13} & -y_3 & -y_1 & y_{13} & -y_3 \\ -y_2 & y_{12} & -y_1 & -y_2 & y_{12} \end{bmatrix}.$$

Weights λ satisfy the conditions of Theorem 1.6.7 provided the following are not zero: $\lambda_1, \lambda_2, \lambda_3, \lambda_4, \lambda_{123}, \lambda_{1234}$. The βnbc basis for $H^2(A^\bullet(\mathcal{T}), a_\lambda)$ is $\{\eta_{24}, \eta_{34}\}$ where $\eta_{ij} = \lambda_i \lambda_j a_{ij}$. The projection $\rho^2 : A^2(\mathcal{T}) \twoheadrightarrow H^2(A^\bullet(\mathcal{T}), a_\lambda)$ is given by

$$\rho^2(a_{ij}) = \begin{cases} \dfrac{\lambda_{12}\eta_{24} + \lambda_2\eta_{34}}{\lambda_1\lambda_2\lambda_{123}} & \text{if } (ij) = (12), \\[2ex] \dfrac{\lambda_3\eta_{24} + \lambda_{13}\eta_{34}}{\lambda_1\lambda_3\lambda_{123}} & \text{if } (ij) = (13), \\[2ex] \dfrac{-(\eta_{24} + \eta_{34})}{\lambda_1\lambda_4} & \text{if } (ij) = (14), \\[2ex] \dfrac{\eta_{ij}}{\lambda_i\lambda_j} & \text{if } (ij) = (24) \text{ or } (34). \end{cases}$$

Calculations with the endomorphisms $\omega_\lambda^2(\mathcal{T}_j, \mathcal{T}) = \omega^2(\mathcal{T}_j, \mathcal{T})\big|_{y \mapsto \lambda}$ and the projection ρ^2 yield the following combinatorial Gauss-Manin connection matrices:

$$\Omega_{\mathcal{C}}^2(\mathcal{T}_1, \mathcal{T}) = \begin{bmatrix} 0 & \lambda_2 \\ 0 & -\lambda_{12} \end{bmatrix}, \quad \Omega_{\mathcal{C}}^2(\mathcal{T}_2, \mathcal{T}) = \begin{bmatrix} \lambda_{12} & \lambda_2 \\ 0 & 0 \end{bmatrix}, \quad \Omega_{\mathcal{C}}^2(\mathcal{T}_3, \mathcal{T}) = \begin{bmatrix} \lambda_{1234} & 0 \\ 0 & \lambda_{1234} \end{bmatrix}.$$

These matrices were first determined by Terao [49] and recovered in [39, Ex. 10.4.2] and [15, Ex. 8.2].

We return to the question of resonant weights in Section 2.5.

1.12 Exercises

Problems 1-8 refer to the following arrangements: Let $(\pm 1, \pm 1, \pm 1)$ be the vertices of the standard cube in $\mathbb{R}^3$. Let $\mathcal{A}_1$ be the set of symmetry planes of the cube. This is a central arrangement. Let $\mathcal{A}_2 = \mathbb{P}\mathcal{A}_1$ be its projective quotient, and let $\mathcal{A}_3 = \mathbf{d}_H\mathcal{A}_2$ be the decone with respect to the plane $H = \ker u_3$.

1) Construct the intersection posets of these arrangements. How are they related?

2) Compute the Poincaré polynomials of these arrangements. How are they related?

3) Compute the β invariants of these arrangements.

4) Find the relations in the Orlik-Solomon algebra for these arrangements. How are they related?

5) Order the hyperplanes in $\mathcal{A}_1$ and find the corresponding **nbc** basis. Do the same for $\mathcal{A}_3$ assuming that the decone was with respect to the largest hyperplane in $\mathcal{A}_1$. What can you conclude?

6) What are the dense edges of these arrangements?

7) Find weights for $\mathcal{A}_3$ which satisfy the conditions of Theorem 1.6.7, and compute the combinatorial cohomology directly.

8) Find nontrivial weights for $\mathcal{A}_3$ which do not satisfy the conditions of Theorem 1.6.7, and compute the combinatorial cohomology directly. Are there several possibilities?

9) Give an example of a nonempty central arrangement $\mathcal{A}$ with $H_1, H_2 \in \mathcal{A}$ so that H_1 is a separator and H_2 is not.

10) Let $\mathcal{A}$ be a nonempty central arrangement with top element $T = \cap_{H \in \mathcal{A}} H$. Define $\mu(\mathcal{A}) = \mu(T)$. Show that $\mu(\mathcal{A}) = -\mu(\mathcal{A}'')$ if H is a separator, and $\mu(\mathcal{A}) = \mu(\mathcal{A}') - \mu(\mathcal{A}'')$ otherwise.

Problems 11-16 refer to the following definition: Let $S = \mathbb{C}[u_1, \ldots, u_\ell]$ be the polynomial ring of V. Let Der_S denote the derivations (vector fields) of V. Given a central arrangement $\mathcal{A}$ defined by the homogeneous polynomial $Q = Q(\mathcal{A})$, define

$$D(\mathcal{A}) = \{\theta \in Der_S \mid \theta(Q) \in QS\}.$$

11) Show that each $\theta \in D(\mathcal{A})$ decomposes into homogeneous components so $D(\mathcal{A})$ is a graded S-module.

12) Show that the Euler derivation $\sum_{i=1}^{\ell} u_i \frac{\partial}{\partial u_i}$ is in $D(\mathcal{A})$ for every nonempty arrangement.

13) Show that if $\mathcal{A} \subset \mathcal{B}$, then $D(\mathcal{B}) \subset D(\mathcal{A})$.

14) Consider $\mathcal{A}_1$ of Problem 1 and compute $D(\mathcal{A}_1)$.

15) Call $\mathcal{A}$ a *free* arrangement if $D(\mathcal{A})$ is a free S-module. Consider $\mathcal{A}_1$ of Problem 1 and show that $\mathcal{A}_1$ is free.

16) Show that $\mathcal{A}$ defined by $Q(\mathcal{A}) = u_1 u_2 u_3 (u_1 + u_2 + u_3)$ is not free.

2

Applications

Much of the algebraic combinatorics described in Chapter 1 was originally developed with topological applications in mind. We give a brief description of some of the main features of these applications.

2.1 Topology

If $\mathcal{A}$ is a real arrangement, then the complement $\mathsf{M}(\mathcal{A})$ consists of open, connected, convex subsets of $\mathbb{R}^\ell$ whose number was determined in Theorem 1.1.3. If the arrangement is essential, then some chambers may be bounded, and the number of bounded chambers was determined in Theorem 1.1.4. Thus from the topological point of view $\mathsf{M}(\mathcal{A})$ is not very interesting.

If $\mathcal{A}$ is a complex arrangement, then the topology of the complement $\mathsf{M}(\mathcal{A})$ is much more subtle. It is an open, connected, smooth manifold of real dimension 2ℓ with interesting homology and homotopy groups. When $\ell = 1$, the complement of n points in $\mathbb{C}$ is homotopy equivalent to a bouquet of n circles. When $\ell \geq 2$, the complement is not easy to visualize. An elementary Morse theoretic argument provides the following:

Proposition 2.1.1. *The complement* $\mathsf{M}(\mathcal{A})$ *has the homotopy type of a finite cell complex.* □

This implies in particular that all homology and cohomology groups are finite dimensional. A much stronger statement was proved by Randell [42] and independently by Dimca and Papadima [23]. Call a cell complex *minimal* if the number of its q-cells equals its q-th betti number for all $q \geq 0$.

Theorem 2.1.2 ([42]). *The complement* $\mathsf{M}(\mathcal{A})$ *is homotopy equivalent to a minimal cell complex.* □

Cohomology Ring

The cohomology ring of the complement was determined by Brieskorn [8]. Let $B(\mathcal{A})^\bullet$ be the graded $\mathbb{C}$-algebra generated by 1 (of degree 0) and the logarithmic forms $d\alpha_k/\alpha_k$ for $H_k \in \mathcal{A}$ (of degree 1).

Theorem 2.1.3 ([8]). *There is an isomorphism of graded algebras*

$$B(\mathcal{A})^\bullet \simeq H^\bullet(M(\mathcal{A}), \mathbb{C})$$

which carries the form $d\alpha_k/\alpha_k$ to the cohomology class which has value $2\pi i$ on the homology class of a simple loop linking H_k, and 0 elsewhere.   □

The Brieskorn algebra $B(\mathcal{A})^\bullet$ is a subalgebra of the de Rham algebra of all rational differential forms on V with poles of arbitrary order on the divisor $N(\mathcal{A})$. If we use the customary total grading whose degree is the sum of the exterior degree and the polynomial degree, then the Brieskorn algebra is the graded component of total degree 0. The Orlik-Solomon algebra, defined in Section 1.4 in purely combinatorial terms, is isomorphic to the Brieskorn algebra and hence to the cohomology algebra.

Theorem 2.1.4 ([37]). *There is an isomorphism of graded algebras*

$$A(\mathcal{A})^\bullet \simeq B(\mathcal{A})^\bullet$$

which carries the generator $a_k \in A(\mathcal{A})^1$ to the form $d\alpha_k/\alpha_k$. In particular, the cohomology algebra of the complement is a combinatorial invariant.   □

Let $b_q(\mathcal{A}) = \dim H^q(M(\mathcal{A}), \mathbb{C})$ be the q-th Betti number of M and let $Poin(M, t) = \sum_{q \geq 0} b_q(\mathcal{A})$ be the Poincaré polynomial of the complement M. In light of Theorem 2.1.3, Theorem 2.1.4, and Corollary 1.4.5 we have:

Corollary 2.1.5. *The Poincaré polynomial of the complement equals the Poincaré polynomial of the arrangement.*

$$Poin(M(\mathcal{A}), t) = \pi(\mathcal{A}, t). \quad \square$$

Homotopy Groups

The fundamental group of the complement, $\pi_1(M, *)$ with respect to a base point $* \in M$ (which will remain unspecified), was first described for complexified real arrangements by Randell [40] then determined by Arvola [5] for all arrangements. A detailed description with proof is also found in [38]. It suffices to consider line arrangements in $\mathbb{C}^2$. The fundamental group of the complement $\pi_1(M, *)$ has a presentation with one generator for each hyperplane chosen as a loop linking the hyperplane. There are two kinds of relators. At each point where hyperplanes intersect, there is a set of relators which involve only the generators corresponding to the hyperplanes which contain the point or their conjugates. There are additional braid relations which depend on the particular wiring diagram used for the presentation. In a complexified real arrangement, the real picture may be used as a wiring diagram. In this case there are no braid relations.

We illustrate this on the Selberg arrangement of Figure 1.4. Tilt the x axis by a small positive angle so that the four points in $L(\mathcal{A})$ have distinct x coordinates: $x_{135} < x_{14} < x_{23} < x_{245}$. A plane P in $\mathbb{C}^2$ whose intersection with the real plane of the figure is a line perpendicular to the x axis intersects the arrangement in five points everywhere except at the four points of $L(\mathcal{A})$. We choose small loops γ_j in P around the five points and label them by the corresponding hyperplanes. At large negative x the order of the hyperplanes with increasing y is H_2, H_1, H_5, H_3, H_4. Now sweep P left to right. As P passes through $H_1 \cap H_3 \cap H_5$, the order of the hyperplanes changes to H_2, H_3, H_5, H_1, H_4. The theory tells us that a set of new relators is added to the group: $[\gamma_3, \gamma_5, \gamma_1]$ defined as

$$\gamma_3 \gamma_5 \gamma_1 = \gamma_1 \gamma_3 \gamma_5 = \gamma_5 \gamma_1 \gamma_3.$$

In addition, some of the loops emerging on the other side have been conjugated. Ordered by increasing y they are $\gamma_2, \gamma_3, \gamma_5^{\gamma_1}, \gamma_1, \gamma_4$, where $\gamma_5^{\gamma_1} = \gamma_1^{-1} \gamma_5 \gamma_1$. Passing through $H_1 \cap H_4$ adds the commutator relation $[\gamma_1, \gamma_4] = \gamma_1 \gamma_4 \gamma_1^{-1} \gamma_4^{-1}$. Passing through the remaining two points of $L(\mathcal{A})$ we obtain the following presentation:

$$\pi_1(\mathsf{M}, *) = \{\gamma_j \mid [\gamma_3, \gamma_5, \gamma_1], [\gamma_1, \gamma_4], [\gamma_2, \gamma_3], [\gamma_4, \gamma_5^{\gamma_1}, \gamma_2]\}.$$

Not much is known about the structure of these groups in general. It follows from the description of the relators that they are elements of the commutator subgroup. Thus the abelianization is free, and the number of generators equals the number of hyperplanes. The following question is over ten years old: is the identity the only element of finite order in $\pi_1(\mathsf{M}, *)$?

Higher homotopy groups are discussed by Falk and Randell [27, 28]. See also Dimca and Papadima [23] for recent developments. Recall that a space is called a $K(\pi, 1)$ space if all its higher homotopy groups vanish. There are examples where $\mathsf{M}(\mathcal{A})$ is a $K(\pi, 1)$ space, such as the complexification of the braid arrangement $\mathcal{B}_\ell$, defined in Section 1.2. There are also examples where $\mathsf{M}(\mathcal{A})$ is not a $K(\pi, 1)$ space, such as the general position arrangements with $n > \ell$, defined in Section 1.1. There are two classes of arrangements whose complements are known to be $K(\pi, 1)$ spaces. A central real arrangement is called *simplicial* if every chamber of the real complement is an open cone over a simplex. Deligne [22] proved that the complement of the complexification of a simplicial real arrangement is a $K(\pi, 1)$ space. The other class is called *fiber type*. Let $\mathcal{A}$ be an ℓ-arrangement. Call $\mathcal{A}$ *strictly linearly fibered* if, after a suitable linear change of coordinates, the restriction of the projection of $\mathsf{M}(\mathcal{A})$ to the first $(\ell - 1)$ coordinates is a fiber bundle projection whose base space B is the complement of an arrangement in $\mathbb{C}^{\ell-1}$ and whose fiber is the complex line $\mathbb{C}$ with finitely many points removed.

Definition 2.1.6. *(1) The 1-arrangement* $(\{0\}, \mathbb{C})$ *is fiber type.*

(2) For $\ell \geq 2$, *the* ℓ-*arrangement* $\mathcal{A}$ *is* fiber type *if* $\mathcal{A}$ *is strictly linearly fibered with base* $B = \mathsf{M}(\mathcal{B})$ *and* $\mathcal{B}$ *is an* $(\ell - 1)$-*arrangement of fiber type.*

The braid arrangement is fiber type. Terao [49] showed that the topological property of fiber type is equivalent to the combinatorial property known as supersolvability of the intersection poset. It follows from repeated application of the the long exact sequence of a fibration that the complement of a fiber type arrangement is a $K(\pi, 1)$ space. We do not know how to determine whether $\mathsf{M}(\mathcal{A})$ is a $K(\pi, 1)$ space in general.

Many interesting invariants have been derived from the fundamantal group. We mention one here, obtained from the *lower central series*. Let $G = \pi_1(\mathsf{M}, *)$, $G_1 = G$, and for $n \geq 1$ define $G_{n+1} = [G_n, G]$ where $[A, B]$ denotes the subgroup generated by commutators of elements in A and B. The groups G_n form the lower central series of G. It is known that $G(n) = G_n/G_{n+1}$ is a finitely generated abelian group, hence its rank ϕ_n is an invariant of G and thus of $\mathcal{A}$. There is considerable literature on finding the ranks ϕ_n. Falk and Randell proved the following.

Theorem 2.1.7 ([26]). *Let* $\mathcal{A}$ *be a fiber type arrangement with complement* M. *Then the following identity holds in* $\mathbb{Z}[[t]]$:

$$\prod_{j=1}^{\infty}(1 - t^j)^{\phi_j} = Poin(\mathsf{M}, -t). \quad \square$$

2.2 Local System Cohomology

Next we turn to connections with the theory of multivariable hypergeometric integrals. Given $\boldsymbol{\lambda}$, define a multivalued holomorphic function on M by

$$\Phi(u; \boldsymbol{\lambda}) = \prod_{i=1}^{n} \alpha_i^{\lambda_i}.$$

A generalized hypergeometric integral is of the form

$$\int_{\sigma} \Phi(u; \boldsymbol{\lambda})\eta$$

where σ is a suitable domain of integration and η is a holomorphic form on M, see [2]. The function $\Phi(u; \boldsymbol{\lambda})$ determines a locally constant sheaf $\mathcal{L}$ on M whose sections are the single valued branches of $\Phi(u; \boldsymbol{\lambda})^{-1}$. The sheaf cohomology groups $H^q(\mathsf{M}; \mathcal{L})$ agree with cohomology in the local system defined below [39, Prop. 2.1.3]. We also defined local system homology groups in [39] and showed how the hypergeometric integrals are evaluations of a pairing between them. In order to study hypergeometric integrals, we investigate these local system cohomology groups.

Stratified Morse Theory

For an arbitrary complex local system $\mathcal{L}$ on the complement of an arrangement $\mathcal{A}$, stratified Morse theory was used in [11] to construct a complex $(K^\bullet(\mathcal{A}), \Delta^\bullet)$, the cohomology of which is naturally isomorphic to $H^\bullet(\mathsf{M}; \mathcal{L})$. We recall this construction in the context of rank one local systems from [11, 13].

Given $\boldsymbol{\lambda}$, let $t_j = \exp(-2\pi i \lambda_j)$ and $\mathbf{t} = (t_1, \ldots, t_n) \in (\mathbb{C}^*)^n$. Associated to $\boldsymbol{\lambda}$ we have a rank one representation $\rho : \pi_1(\mathsf{M}) \to \mathbb{C}^*$, given by $\rho(\gamma_j) = t_j$, where γ_j is any meridian loop about the hyperplane H_j of $\mathcal{A}$, and a corresponding rank one local system $\mathcal{L} = \mathcal{L}_{\mathbf{t}} = \mathcal{L}_{\boldsymbol{\lambda}}$ on M. Note that weights $\boldsymbol{\lambda}$ and $\boldsymbol{\lambda}'$ yield identical representations and local systems if $\boldsymbol{\lambda} - \boldsymbol{\lambda}' \in \mathbb{Z}^n$.

We assume throughout that $\mathcal{A}$ contains ℓ linearly independent hyperplanes. Then $\mathcal{A}$ determines a stratification of V with strata $S_X = \mathsf{M}(\mathcal{A}^X)$ for $X \in L(\mathcal{A})$. Let $\mathcal{F}$ be a complete flag (of affine subspaces) in $\mathbb{C}^\ell$,

$$\mathcal{F}: \quad \emptyset = \mathcal{F}^{-1} \subset \mathcal{F}^0 \subset \mathcal{F}^1 \subset \mathcal{F}^2 \subset \cdots \subset \mathcal{F}^\ell = \mathbb{C}^\ell,$$

transverse to the stratification determined by $\mathcal{A}$, so that $\dim \mathcal{F}^q \cap S_X = q - \operatorname{codim} S_X$ for each stratum, where a negative dimension indicates that $\mathcal{F}^q \cap S_X = \emptyset$. For an explicit construction of such a flag, see [11, §1]. Let $\mathsf{M}^q = \mathcal{F}^q \cap \mathsf{M}$ for each q. Let $K^q = H^q(\mathsf{M}^q, \mathsf{M}^{q-1}; \mathcal{L})$, and denote by Δ^q the boundary homomorphism $H^q(\mathsf{M}^q, \mathsf{M}^{q-1}; \mathcal{L}) \to H^{q+1}(\mathsf{M}^{q+1}, \mathsf{M}^q; \mathcal{L})$ of the triple $(\mathsf{M}^{q+1}, \mathsf{M}^q, \mathsf{M}^{q-1})$. The following compiles several results from [11].

Theorem 2.2.1. *Let $\mathcal{L}$ be a complex rank one local system on* M.

1. *For each q, $0 \le q \le \ell$, we have $H^i(\mathsf{M}^q, \mathsf{M}^{q-1}; \mathcal{L}) = 0$ if $i \ne q$, and $\dim_{\mathbb{C}} H^q(\mathsf{M}^q, \mathsf{M}^{q-1}; \mathcal{L}) = b_q(\mathcal{A})$ is equal to the q-th Betti number of* M *with trivial local coefficients $\mathbb{C}$.*

2. *The system of complex vector spaces and linear maps $(K^\bullet, \Delta^\bullet)$,*

$$K^0 \xrightarrow{\Delta^0} K^1 \xrightarrow{\Delta^1} K^2 \longrightarrow \cdots \longrightarrow K^{\ell-1} \xrightarrow{\Delta^{\ell-1}} K^\ell,$$

is a complex ($\Delta^{q+1} \circ \Delta^q = 0$). The cohomology of this complex is naturally isomorphic to $H^\bullet(\mathsf{M}; \mathcal{L})$, the cohomology of M *with coefficients in $\mathcal{L}$.* □

Universal Complex

The dimensions of the terms K^q of the complex $(K^\bullet, \Delta^\bullet)$ are independent of $\mathbf{t}$ (resp., $\boldsymbol{\lambda}$, $\mathcal{L}$). Write $\Delta^\bullet = \Delta^\bullet(\mathbf{t})$ to indicate the dependence of the complex on $\mathbf{t}$, and view these boundary maps as functions of $\mathbf{t}$. Let $\Lambda = \mathbb{C}[x_1^{\pm 1}, \ldots, x_n^{\pm 1}]$ be the ring of complex Laurent polynomials in n commuting variables, and for each q, let $\mathsf{K}^q = \Lambda \otimes_{\mathbb{C}} K^q$.

Theorem 2.2.2 ([13, Thm. 2.9]). *For an arrangement $\mathcal{A}$ of n hyperplanes with complement* M *there exists a universal complex $(\mathsf{K}^\bullet, \Delta^\bullet(\mathbf{x}))$ with the following properties:*

1. *The terms are free Λ-modules, whose ranks are given by the Betti numbers of* M, $K^q \simeq \Lambda^{b_q(\mathcal{A})}$.
2. *The boundary maps $\Delta^q(\mathbf{x}) : K^q \to K^{q+1}$ are Λ-linear.*
3. *For each $\mathbf{t} \in (\mathbb{C}^*)^n$ the specialization $\mathbf{x} \mapsto \mathbf{t}$ yields the complex $(K^\bullet, \Delta^\bullet(\mathbf{t}))$, the cohomology of which is isomorphic to $H^\bullet(M; \mathcal{L}_\mathbf{t})$, the cohomology of* M *with coefficients in the local system associated to $\mathbf{t}$.*   □

The entries of the boundary maps $\Delta^q(\mathbf{x})$ are elements of the Laurent polynomial ring Λ, the coordinate ring of the complex algebraic n-torus. Via the specialization $\mathbf{x} \mapsto \mathbf{t} \in (\mathbb{C}^*)^n$, we view them as holomorphic functions $(\mathbb{C}^*)^n \to \mathbb{C}$. Thus we may view $\Delta^q(\mathbf{x})$ as a holomorphic map $\Delta^q : (\mathbb{C}^*)^n \to \mathrm{Mat}(\mathbb{C})$, $\mathbf{t} \mapsto \Delta^q(\mathbf{t})$ from the complex torus to matrices with complex entries.

Remark 2.2.3. The canonical graded algebra isomorphism $H^\bullet(M, \mathbb{C}) \simeq A^\bullet(\mathcal{A})$ induces an isomorphism of vector spaces $K^q(\mathcal{A}) \simeq A^q(\mathcal{A})$ for $0 \le q \le \ell$. This isomorphism is not canonical.

This universal complex is closely related to the Aomoto complex.

Theorem 2.2.4 ([13]). *For any arrangement $\mathcal{A}$, the Aomoto complex $(A^\bullet, a_\mathbf{y})$ is chain equivalent to the linearization of the universal complex $(K^\bullet, \Delta^\bullet(\mathbf{x}))$ at $\mathbf{x} = 1$.*   □

Nonresonant Weights

The complex $(K^\bullet, \Delta^\bullet(\mathbf{t}))$ computes local system cohomology in principle, but we do not know how to calculate the groups $H^q(M; \mathcal{L})$ explicitly for arbitrary weights. It is clear that for a dense open set of weights the linearized complex has the same cohomology groups as the universal complex. We call such weights *nonresonant*. A sufficient condition for this was established by Esnault-Schechtman-Varchenko [24] and improved by Schechtman-Terao-Varchenko:

Theorem 2.2.5 ([44]). *Let $\mathcal{A}$ be an arrangement of rank ℓ with projective closure $\mathcal{A}_\infty$. Assume that $\lambda_X \notin \mathbb{Z}_{>0}$ for every $X \in D(\mathcal{A}_\infty)$. Then*

$$H^q(M; \mathcal{L}) \simeq H^q(A^\bullet, a_\lambda).   \square$$

Theorem 2.2.5 was obtained using the de Rham complex, resolution of singularities, and work of Deligne [21]. It would be interesting to obtain an independent argument via stratified Morse theory.

We combine Corollary 1.6.5 with Theorem 2.2.5 to obtain the following result.

Theorem 2.2.6 ([44]). *Let* M *be the complement of an essential arrangement $\mathcal{A}$ in $\mathbb{C}^\ell$. If $\mathcal{L}$ is a rank one local system on* M *whose weights λ satisfy the condition*

(STV) $\lambda_X \notin \mathbb{Z}_{\ge 0}$ *for every dense edge X of $\mathcal{A}_\infty$,*

then $H^q(M; \mathcal{L}) = 0$ for $q \ne \ell$ and $\dim H^\ell(M; \mathcal{L}) = \beta(\mathcal{A})$.   □

2.3 Resonance

Consider the set of weights $\lambda \in \mathbb{C}^n$. For nonresonant weights we have $H^q(\mathsf{M}; \mathcal{L}) = 0$ for $q \neq \ell$ and $\dim H^\ell(\mathsf{M}; \mathcal{L}) = \beta(\mathcal{A})$. It follows from Theorem 1.6.3 that the (STV) condition depends only on the combinatorial type $\mathcal{T}$. We call weights which satisfy the (STV) condition $\mathcal{T}$-*nonresonant*. The set of $\mathcal{T}$-nonresonant weights is a proper subset of all nonresonant weights. For example, in case of general position arrangements, $\mathcal{G}$-nonresonance requires $\lambda_H \notin \mathbb{Z}_{\geq 0}$ for all $H \in \mathcal{A}_\infty$, while every nontrivial local system is nonresonant. Note also that $\mathcal{T}$-nonresonant weights form a dense, open subset of all weights.

We call λ *combinatorial* if $H^q(\mathsf{M}; \mathcal{L}) \simeq H^q(A^\bullet, a_\lambda)$ for all q. We say that λ is $\mathcal{T}$-*combinatorial* if there exists λ' so that $\lambda - \lambda' \in \mathbb{Z}^n$ and λ' satisfies the conditions of Theorem 2.2.5. The corresponding sets are dense, open subsets of all weights and contain both nonresonant and resonant weights. See Exercise 3 for an example. There are also examples of arrangements where $H^q(\mathsf{M}; \mathcal{L}) \not\simeq H^q(A^\bullet, a_\lambda)$. Exercise 4 provides an arrangement and weights for which $\dim H^1(A^\bullet, a_\lambda) = 1$ and $\dim H^1(\mathsf{M}; \mathcal{L}) = 2$, see [20].

There is no general algorithm for calculating local system cohomology for resonant weights. It is an interesting question to determine the structure of the subset of resonant weights in the space of all weights. Each point $\mathbf{t} \in (\mathbb{C}^*)^n$ gives rise to a local system $\mathcal{L} = \mathcal{L}_\mathbf{t}$ on the complement M. For nonresonant $\mathbf{t}$, the cohomology $H^q(\mathsf{M}, \mathcal{L}_\mathbf{t})$ vanishes for $q < \ell$. Those $\mathbf{t}$ for which $H^q(\mathsf{M}; \mathcal{L}_\mathbf{t})$ does not vanish comprise the *characteristic varieties*

$$\Sigma_m^q(\mathsf{M}) = \{\mathbf{t} \in (\mathbb{C}^*)^n \mid \dim H^q(\mathsf{M}; \mathcal{L}_\mathbf{t}) \geq m\}.$$

These loci are algebraic subvarieties of $(\mathbb{C}^*)^n$, which are invariants of the homotopy type of M. See Arapura [3] and Libgober [34] for detailed discussions of these varieties in the contexts of quasiprojective varieties and plane algebraic curves. The characteristic varieties are closely related to the resonance varieties defined in Section 1.6.

Theorem 2.3.1 ([13]). *Let $\mathcal{A}$ be an arrangement in $\mathbb{C}^\ell$ with complement M and Orlik-Solomon algebra A. For each q and m, the resonance variety $\mathcal{R}_m^q(A)$ coincides with the tangent cone of the characteristic variety $\Sigma_m^q(\mathsf{M})$ at the point $\mathbf{1} = (1, \ldots, 1) \in (\mathbb{C}^*)^n$.*

Proof. For each $\mathbf{t} \in (\mathbb{C}^*)^n$, the cohomology of M with coefficients in the local system $\mathcal{L}_\mathbf{t}$ is isomorphic to that of the Morse theoretic complex $(K^\bullet(\mathcal{A}), \Delta^\bullet(\mathbf{t}))$, the specialization at $\mathbf{t}$ of the universal complex $(K_A^\bullet(\mathcal{A}), \Delta^\bullet(\mathbf{x}))$ of Theorem 2.2.2. So $\mathbf{t} \in \Sigma_m^q(\mathsf{M})$ if and only if $\dim H^q(K^\bullet(\mathcal{A}), \Delta^\bullet(\mathbf{t})) \geq m$. An exercise in linear algebra shows that

$$\Sigma_m^q(\mathsf{M}) = \{\mathbf{t} \in (\mathbb{C}^*)^n \mid \operatorname{rank} \Delta^{q-1}(\mathbf{t}) + \operatorname{rank} \Delta^q(\mathbf{t}) \leq \dim K^q(\mathcal{A}) - m\}.$$

For $\lambda \in \mathbb{C}^n$, we have $\lambda \in \mathcal{R}_m^q$ if $\dim H^q(A^\bullet, a_\lambda) \geq m$. Denote the matrix of $a_\lambda \wedge : A^q(\mathcal{A}) \to A^{q+1}(\mathcal{A})$ by $\mu^q(\lambda)$. Then, as above,

$$\mathcal{R}_m^q(A) = \{\boldsymbol{\lambda} \in \mathbb{C}^n \mid \operatorname{rank} \mu^{q-1}(\boldsymbol{\lambda}) + \operatorname{rank} \mu^q(\boldsymbol{\lambda}) \le \dim A^q(\mathcal{A}) - m\}.$$

Now $\dim A^q(\mathcal{A}) = \dim K^q(\mathcal{A}) = b_q(\mathcal{A})$, and for each q, by Theorem 2.2.4, we have $\operatorname{rank} \mu^q(\boldsymbol{\lambda}) = \operatorname{rank} \Delta_*^q(\boldsymbol{\lambda})$ where $\Delta_*^q(\boldsymbol{\lambda})$ denotes the linearization of $\Delta^q(\boldsymbol{\lambda})$. Thus,

$$\Sigma_m^q(\mathsf{M}) = \{\mathbf{t} \in (\mathbb{C}^*)^n \mid \operatorname{rank} \Delta^{q-1}(\mathbf{t}) + \operatorname{rank} \Delta^q(\mathbf{t}) \le b_q(\mathcal{A}) - m\} \text{ and}$$
$$\mathcal{R}_m^q(A) = \{\boldsymbol{\lambda} \in \mathbb{C}^n \mid \operatorname{rank} \Delta_*^{q-1}(\boldsymbol{\lambda}) + \operatorname{rank} \Delta_*^q(\boldsymbol{\lambda}) \le b_q(\mathcal{A}) - m\},$$

and the result follows. □

The characteristic varieties are known to be unions of torsion-translated subtori of $(\mathbb{C}^*)^n$, see [3]. In particular, all irreducible components of $\Sigma_m^q(\mathsf{M})$ passing through $\mathbf{1}$ are subtori of $(\mathbb{C}^*)^n$. Consequently, all irreducible components of the tangent cone are linear subspaces of $\mathbb{C}^n$.

Corollary 2.3.2. *For each q and m, the resonance variety $\mathcal{R}_m^q(A)$ is the union of an arrangement of subspaces in $\mathbb{C}^n$.* □

For $q = 1$, these results were established by Cohen and Suciu [20], see also Libgober and Yuzvinsky [34, 36]. For the discriminantal arrangements of Schechtman and Varchenko [46], they were established in [12]. In particular, as conjectured by Falk [25, Conjecture 4.7], the resonance varieties $\mathcal{R}_m^q(A)$ were known to be unions of linear subspaces in these instances. Corollary 2.3.2 above resolves this conjecture positively for all arrangements in all dimensions. Theorem 2.3.1 and Corollary 2.3.2 have been obtained by Libgober in a more general situation, see [35].

There are examples of arrangements where the characteristic varieties contain (positive dimensional) components which do not pass through $\mathbf{1}$ and hence cannot be detected by the resonance variety, see Suciu [48]. In some of these cases the combinatorial cohomology vanishes but the local system cohomology does not.

2.4 Moduli Spaces

Recall from Section 1.7 that the moduli space of all multi-arrangements of n ordered hyperplanes in $\mathbb{C}^\ell$ may be viewed as the set of matrices (1.11) whose rows are elements of $\mathbb{CP}^\ell$, and that given a realizable type $\mathcal{T}$, the moduli space of type $\mathcal{T}$ is

$$\mathsf{X}(\mathcal{T}) = \{\mathbf{b} \in (\mathbb{CP}^\ell)^n \mid \Delta_J(\mathbf{b}) = 0 \text{ for } J \in \operatorname{Dep}(\mathcal{T})_{\ell+1}, \ \Delta_J(\mathbf{b}) \ne 0 \text{ else}\}.$$

Let

$$\mathsf{Y}(\mathcal{T}) = \{\mathbf{b} \in (\mathbb{CP}^\ell)^n \mid \Delta_J(\mathbf{b}) \ne 0 \text{ for } J \notin \operatorname{Dep}(\mathcal{T})_{\ell+1}\}.$$

Then the moduli space of type $\mathcal{T}$ may be realized as

$$X(\mathcal{T}) = \{\mathsf{b} \in \mathsf{Y}(\mathcal{T}) \mid \Delta_J(\mathsf{b}) = 0 \text{ for } J \in \mathrm{Dep}(\mathcal{T})_{\ell+1}\}.$$

Note that $\mathsf{X}(\mathcal{G}) = \mathsf{Y}(\mathcal{G})$. For any other type $\mathcal{T}$ the moduli space $\mathsf{X}(\mathcal{G})$ may be realized as

$$\mathsf{X}(\mathcal{G}) = \{\mathsf{b} \in \mathsf{Y}(\mathcal{T}) \mid \Delta_J(\mathsf{b}) \neq 0 \text{ for } J \in \mathrm{Dep}(\mathcal{T})_{\ell+1}\}.$$

If $\mathcal{T} \neq \mathcal{G}$, then $\mathsf{X}(\mathcal{T})$ and $\mathsf{X}(\mathcal{G})$ are disjoint subspaces of $\mathsf{Y}(\mathcal{T})$. Let $i_{\mathcal{T}} : \mathsf{X}(\mathcal{G}) \to \mathsf{Y}(\mathcal{T})$ and $j_{\mathcal{T}} : \mathsf{X}(\mathcal{T}) \to \mathsf{Y}(\mathcal{T})$ denote the natural inclusions. We showed in [15] that for any combinatorial type $\mathcal{T}$ the inclusion $i_{\mathcal{T}} : \mathsf{X}(\mathcal{G}) \to \mathsf{Y}(\mathcal{T})$ induces a surjection $(i_{\mathcal{T}})_* : H_1(\mathsf{X}(\mathcal{G})) \to H_1(\mathsf{Y}(\mathcal{T}))$.

For the type $\mathcal{G}$ of general position arrangements, the closure of the moduli space is $\overline{\mathsf{X}}(\mathcal{G}) = (\mathbb{CP}^\ell)^n$. The divisor $\mathsf{D}(\mathcal{G}) = \overline{\mathsf{X}}(\mathcal{G}) \setminus \mathsf{X}(\mathcal{G})$ is given by $\mathsf{D}(\mathcal{G}) = \bigcup_J \mathsf{D}_J$, whose components $\mathsf{D}_J = \{\mathsf{b} \in (\mathbb{CP}^\ell)^n \mid \Delta_J(\mathsf{b}) = 0\}$ are irreducible hypersurfaces indexed by $J = \{j_1, \ldots, j_{\ell+1}\}$.

Choose a basepoint $\mathbf{c} \in \mathsf{X}(\mathcal{G})$, and for each $\ell+1$ element subset J of $[n+1]$, let $\mathbf{d}_J$ be a generic point in D_J. Let $\varGamma_J$ be a meridian loop based at $\mathbf{c}$ in $\mathsf{X}(\mathcal{G})$ about the point $\mathbf{d}_J \in \mathsf{D}_J$. Note that $\mathbf{c} \in \mathsf{Y}(\mathcal{T})$ and that $\varGamma_J$ is a (possibly null-homotopic) loop in $\mathsf{Y}(\mathcal{T})$ for any combinatorial type $\mathcal{T}$. We showed in [15] that for any combinatorial type $\mathcal{T}$ the homology group $H_1(\mathsf{Y}(\mathcal{T}))$ is generated by the classes $\{[\varGamma_J] \mid J \notin \mathrm{Dep}(\mathcal{T})_{\ell+1}\}$. Thus the homology group $H_1(\mathsf{X}(\mathcal{G}))$ is generated by the classes $[\varGamma_J]$ where J ranges over all $\ell+1$ element subsets of $[n+1]$.

It is easy to see that the moduli space $\mathsf{X}(\mathcal{T}')$ has complex codimension one in the closure $\overline{\mathsf{X}}(\mathcal{T})$ if and only if $\mathcal{T}$ covers $\mathcal{T}'$. The next theorem is essential for later results.

Theorem 2.4.1 ([15]). *Let $\mathcal{T}$ be a combinatorial type which covers the type $\mathcal{T}'$. Let b' be a point in $\mathsf{X}(\mathcal{T}')$, and $\gamma \in \pi_1(\mathsf{X}(\mathcal{T}), \mathsf{b})$ a simple loop in $\mathsf{X}(\mathcal{T})$ about b'. Then the homology class $[\gamma]$ satisfies*

$$(j_{\mathcal{T}})_*([\gamma]) = \sum_{J \in \mathrm{Dep}(\mathcal{T}', \mathcal{T})} m_J \cdot [\varGamma_J] \tag{2.1}$$

where m_J is the order of vanishing of the restriction of Δ_J to $\overline{\mathsf{X}}(\mathcal{T})$ along $\mathsf{X}(\mathcal{T}')$. □

It is not hard to see that this multiplicity agrees with the algebraic definition given in Section 1.9.

The moduli space $\mathsf{X}(\mathcal{T})$ is not necessarily connected. The existence of a combinatorial type whose moduli space has at least two components follows from examples of Rybnikov [43, 4]. Let $\mathsf{B}(\mathcal{T})$ be a smooth component of the moduli space. Corresponding to each $\mathsf{b} \in \mathsf{B}(\mathcal{T})$ we have an arrangement $\mathcal{A}_\mathsf{b}$, combinatorially equivalent to $\mathcal{A}$, with hyperplanes defined by the first n rows of the matrix equation $\mathsf{b} \cdot \tilde{\mathsf{u}} = 0$, where $\tilde{\mathsf{u}} = \begin{pmatrix} 1 & u_1 & \cdots & u_\ell \end{pmatrix}^\top$. Let $\mathsf{M}_\mathsf{b} = M(\mathcal{A}_\mathsf{b})$ be the complement of $\mathcal{A}_\mathsf{b}$. Let

$$M(\mathcal{T}) = \{(\mathsf{b}, \mathsf{u}) \in (\mathbb{CP}^\ell)^n \times \mathbb{C}^\ell \mid \mathsf{b} \in \mathsf{B}(\mathcal{T}) \text{ and } \mathsf{u} \in \mathsf{M_b}\},$$

and define $\pi_{\mathcal{T}} : M(\mathcal{T}) \to \mathsf{B}(\mathcal{T})$ by $\pi_{\mathcal{T}}(\mathsf{b}, \mathsf{u}) = \mathsf{b}$. Since $\mathsf{B}(\mathcal{T})$ is connected by assumption, a result of Randell [41] implies that $\pi_{\mathcal{T}} : M(\mathcal{T}) \to \mathsf{B}(\mathcal{T})$ is a bundle with fiber $\pi_{\mathcal{T}}^{-1}(\mathsf{b}) = \mathsf{M_b}$.

Since this fiber bundle is locally trivial, there is an associated flat vector bundle $\pi^q : \mathbf{H}^q(A) \to \mathsf{B}(\mathcal{T})$ with fiber the combinatorial cohomology group $(\pi^q)^{-1}(\mathsf{b}) = H^q(A(\mathcal{A}_\mathsf{b}), a_\lambda)$ at $\mathsf{b} \in \mathsf{B}(\mathcal{T})$ for each q, $0 \leq q \leq \ell$. Fixing a basepoint $\mathsf{b} \in \mathsf{B}(\mathcal{T})$, the operation of parallel translation of fibers over curves in $\mathsf{B}(\mathcal{T})$ provides a complex representation

$$\Phi^q : \pi_1(\mathsf{B}(\mathcal{T}), \mathsf{b}) \longrightarrow \mathrm{Aut}_{\mathbb{C}}(H^q(A^\bullet(\mathcal{T}), a_\lambda)). \tag{2.2}$$

Clearly, this representation is induced by the representation on the cochain level:

$$\phi^\bullet : \pi_1(\mathsf{B}(\mathcal{T}), \mathsf{b}) \longrightarrow \mathrm{Aut}_{\mathbb{C}}(A^\bullet(\mathcal{T})). \tag{2.3}$$

In both cases we subscript by $\mathcal{T}$ to indicate the dependence of this representation on the combinatorial type $\mathcal{T}$. Since $\mathsf{X}(\mathcal{G})$ is connected, we may write $\mathsf{B}(\mathcal{G})$ in its place. Results in [15] provide a means for comparing the representations

$$\phi_{\mathcal{T}}^\bullet : \pi_1(\mathsf{B}(\mathcal{T}), \mathsf{b}) \to \mathrm{Aut}_{\mathbb{C}}(A^\bullet(\mathcal{T})) \quad \text{and} \quad \phi_{\mathcal{G}}^\bullet : \pi_1(\mathsf{B}(\mathcal{G}), \mathsf{c}) \to \mathrm{Aut}_{\mathbb{C}}(A^\bullet(\mathcal{G})),$$

corresponding to an arbitrary arrangement of n hyperplanes in $\mathbb{C}^\ell$ and a general position arrangement of n hyperplanes in $\mathbb{C}^\ell$.

Theorem 2.4.2 ([15]). *Let $\mathcal{T}$ be a combinatorial type whose first $\ell + 1$ hyperplanes are linearly independent. Let $s : (A(\mathcal{G}), a_\lambda) \to (A(\mathcal{T}), a_\lambda)$ be the natural surjection defined by the respective **nbc** bases. Then for any element $\gamma \in \pi_1(\mathsf{B}(\mathcal{T}), \mathsf{b})$, there is an element $\Gamma \in \pi_1(\mathsf{B}(\mathcal{G}), \mathsf{c})$ so that the diagram*

$$
\begin{array}{ccc}
A^\bullet(\mathcal{G}) & \xrightarrow{\ s\ } & A^\bullet(\mathcal{T}) \\
{\scriptstyle \phi_{\mathcal{G}}^\bullet(\Gamma)}\downarrow & & \downarrow{\scriptstyle \phi_{\mathcal{T}}^\bullet(\gamma)} \\
A^\bullet(\mathcal{G}) & \xrightarrow{\ s\ } & A^\bullet(\mathcal{T})
\end{array}
$$

commutes up to conjugacy. $\qquad\qquad\qquad\qquad\qquad\qquad\qquad\qquad\square$

The requirement is that if $\gamma \in \pi_1(\mathsf{B}(\mathcal{T}), \mathsf{b})$, then $\Gamma \in \pi_1(\mathsf{B}(\mathcal{G}), \mathsf{c})$ satisfies $(i_{\mathcal{T}})_*([\Gamma]) = (j_{\mathcal{T}})_*([\gamma])$. We call such a pair of classes *compatible*.

2.5 Gauss-Manin Connections

In this section we derive the main topological application of the formal connections constructed in Chapter 1.

For each $\mathsf{b} \in \mathsf{B}(\mathcal{T})$, weights $\boldsymbol{\lambda}$ define a local system $\mathcal{L}_\mathsf{b}$ on $\mathsf{M_b}$. Since $\pi_{\mathcal{T}} : M(\mathcal{T}) \to \mathsf{B}(\mathcal{T})$ is locally trivial, there is an associated flat vector bundle

$\pi^q : \mathbf{H}^q(\mathcal{L}) \to \mathsf{B}(\mathcal{T})$ with fiber the local system cohomology group $(\pi^q)_{\mathcal{L}}^{-1}(\mathsf{b}) = H^q(\mathsf{M_b}; \mathcal{L_b})$ at $\mathsf{b} \in \mathsf{B}(\mathcal{T})$ for each q, $0 \le q \le \ell$. Fixing a basepoint $\mathsf{b} \in \mathsf{B}(\mathcal{T})$, the operation of parallel translation of fibers over curves in $\mathsf{B}(\mathcal{T})$ provides a complex representation

$$\Psi_{\mathcal{T}}^q : \pi_1(\mathsf{B}(\mathcal{T}), \mathsf{b}) \longrightarrow \mathrm{Aut}_{\mathbb{C}}(H^q(\mathsf{M_b}; \mathcal{L_b})). \tag{2.4}$$

This representation is related to the corresponding representation of the fundamental group of a general position arrangement of n hyperplanes in $\mathbb{C}^\ell$ with total space $\mathsf{E}^{n,\ell}$ in case of $\mathcal{T}$-nonresonant weights by the following result. Recall from Theorem 2.2.6 that here only $q = \ell$ matters.

Theorem 2.5.1 ([15]). *Let* $\boldsymbol{\lambda}$ *be a collection of* $\mathcal{T}$*-nonresonant weights. Assume that* $\gamma \in \pi_1(\mathsf{B}(\mathcal{T}), \mathsf{b})$ *and* $\Gamma \in \pi_1(\mathsf{B}(\mathcal{G}), \mathbf{c})$ *are compatible classes. Then the automorphism* $\Psi_{\mathcal{T}}^\ell(\gamma) \in \mathrm{Aut}_{\mathbb{C}}(H^\ell(\mathsf{M_b}; \mathcal{L_b}))$ *is determined by* $\Psi_{\mathcal{G}}^\ell(\Gamma) \in \mathrm{Aut}_{\mathbb{C}}(H^\ell(\mathsf{E_c}^{n,\ell}; \mathcal{L_c}))$ *and the surjection* $\bar{s}_{\mathsf{b}}^\ell : H^\ell(\mathsf{E_c}^{n,\ell}; \mathcal{L_c}) \twoheadrightarrow H^\ell(\mathsf{M_b}; \mathcal{L_b})$ *up to conjugacy.* □

The local system cohomology of $\mathsf{M_b}$ may be computed using the Morse theoretic complex $K^\bullet(\mathcal{A}_\mathsf{b})$. The fundmental group of $\mathsf{B}(\mathcal{T})$ acts by chain automorphisms on this complex, see [14, Cor. 3.2], yielding a representation

$$\psi_{\mathcal{T}}^\bullet : \pi_1(\mathsf{B}(\mathcal{T}), \mathsf{b}) \longrightarrow \mathrm{Aut}_{\mathbb{C}}(K^\bullet(\mathcal{A}_\mathsf{b})).$$

Theorem 2.5.2. *The representation* $\Psi_{\mathcal{T}}^q : \pi_1(\mathsf{B}(\mathcal{T}), \mathsf{b}) \to \mathrm{Aut}_{\mathbb{C}}(H^q(\mathsf{M_b}; \mathcal{L_b}))$ *is induced by the representation* $\psi_{\mathcal{T}}^\bullet : \pi_1(\mathsf{B}(\mathcal{T}), \mathsf{b}) \to \mathrm{Aut}_{\mathbb{C}}(K^\bullet(\mathcal{A}_\mathsf{b}))$. □

Bundles and Connections

The vector bundle $\pi^q : \mathbf{H}^q(\mathcal{L}) \to \mathsf{B}(\mathcal{T})$ supports a Gauss-Manin connection corresponding to the representation (2.4). Over a manifold X there is a well known equivalence between local systems and complex vector bundles equipped with flat connections, see [21, 33]. Let $\mathbf{V} \to X$ be such a bundle, with connection ∇. The latter is a $\mathbb{C}$-linear map $\nabla : \mathcal{E}^0(\mathbf{V}) \to \mathcal{E}^1(\mathbf{V})$ where $\mathcal{E}^p(\mathbf{V})$ denotes the complex p-forms on X with values in $\mathbf{V}$, which satisfies $\nabla(f\sigma) = \sigma df + f\nabla(\sigma)$ for a function f and $\sigma \in \mathcal{E}^0(\mathbf{V})$. The connection extends to a map $\nabla : \mathcal{E}^p(\mathbf{V}) \to \mathcal{E}^{p+1}(\mathbf{V})$ for $p \ge 0$, and is flat if the curvature $\nabla \circ \nabla$ vanishes. Call two connections ∇ and ∇' on $\mathbf{V}$ isomorphic if ∇' is obtained from ∇ by a gauge transformation, $\nabla' = g \circ \nabla \circ g^{-1}$ for some $g : X \to \mathrm{Hom}(\mathbf{V}, \mathbf{V})$.

The aforementioned equivalence is given by $(\mathbf{V}, \nabla) \mapsto \mathbf{V}^\nabla$ where $\mathbf{V}^\nabla$ is the local system, or locally constant sheaf, of horizontal sections $\{\sigma \in \mathcal{E}^0(\mathbf{V}) \mid \nabla(\sigma) = 0\}$. There is also a well known equivalence between local systems on X and finite dimensional complex representations of the fundamental group of X. Note that isomorphic connections give rise to the same representation. Under these equivalences, the local system on $X = \mathsf{B}(\mathcal{T})$ induced by the representation Ψ_T^q corresponds to a flat connection on the vector bundle $\pi^q :$ $\mathbf{H}^q(\mathcal{L}) \to \mathsf{B}(T)$, the Gauss-Manin connection.

Let $\gamma \in \pi_1(\mathsf{B}(T), \mathsf{b})$, and let $g : \mathbb{S}^1 \to \mathsf{B}(T)$ be a representative loop. Pulling back the bundle $\pi^q : \mathbf{H}^q(\mathcal{L}) \to \mathsf{B}(T)$ and the Gauss-Manin connection ∇, we obtain a flat connection $g^*(\nabla)$ on the vector bundle over the circle corresponding to the representation of $\pi_1(\mathbb{S}^1, 1) = \langle \zeta \rangle = \mathbb{Z}$ given by $\zeta \mapsto \Psi_T^q(\gamma)$. This vector bundle is trivial, since any map from the circle to the relevant classifying space is null-homotopic. Specifying the flat connection $g^*(\nabla)$ amounts to choosing a logarithm of $\Psi_T^q(\gamma)$. The connection $g^*(\nabla)$ is determined by a connection 1-form $dz/z \otimes \Omega_T^q(\gamma)$ where the connection matrix $\Omega_T^q(\gamma)$ corresponding to γ satisfies $\Psi_T^q(\gamma) = \exp(-2\pi i \, \Omega_T^q(\gamma))$. If γ and $\hat{\gamma}$ are conjugate in $\pi_1(\mathsf{B}(T), \mathsf{b})$, then the resulting connection matrices are conjugate, and the corresponding connections on the trivial vector bundle over the circle are isomorphic. In this sense, the connection matrix $\Omega_T^q(\gamma)$ is determined by the homology class $[\gamma]$ of γ.

In the special case when T covers T' and $\gamma \in \pi_1(\mathsf{B}(T), \mathsf{b})$ is a simple loop linking $\mathsf{B}(T')$ in $\overline{\mathsf{B}(T)}$, we denote the corresponding Gauss-Manin connection in the bundle $\pi^q : \mathbf{H}^q(\mathcal{L}) \to \mathsf{B}(T)$ by $\Omega_{\mathcal{L}}^q(\mathsf{B}(T'), \mathsf{B}(T))$.

Combinatorial Analogs

The Gauss-Manin connection in local system cohomology has combinatorial analogs. We have the vector bundle $\mathbf{A}^q \to \mathsf{B}(T)$ whose fiber at b is $A^q(\mathcal{A}_\mathsf{b})$, the q-th graded component of the Orlik-Solomon algebra of the arrangement $\mathcal{A}_\mathsf{b}$. The **nbc** basis provides a global trivialization of this bundle. Given weights $\boldsymbol{\lambda}$, the cohomology of the complex $(A^\bullet(\mathcal{A}_\mathsf{b}), a_{\boldsymbol{\lambda}})$ gives rise to the flat vector bundle $\mathbf{H}^q(A) \to \mathsf{B}(T)$ whose fiber at b is the q-th combinatorial cohomology group $H^q(A^\bullet(\mathcal{A}_\mathsf{b}), a_{\boldsymbol{\lambda}})$ described above. Like their topological counterparts, these algebraic vector bundles admit flat connections. We call the connection in this cohomology bundle $\Omega_A^q(\mathsf{B}(T'), \mathsf{B}(T))$ temporarily, and show in the next theorem that it is equal to the combinatorial Gauss-Manin connection $\Omega_{\mathcal{C}}^q(T', T)$, defined in Section 1.10.

Theorem 2.5.3. *Let* M *be the complement of an arrangement of type T and let $\mathcal{L}$ be the local system on* M *defined by weights $\boldsymbol{\lambda}$. Suppose T covers T'.*

1. *Let $\rho^q : A^q \twoheadrightarrow H^q(A^\bullet(T), a_{\boldsymbol{\lambda}})$ be the natural projection. Then the connection $\Omega_A^q(\mathsf{B}(T'), \mathsf{B}(T))$ is determined by the equation*

$$\rho^q \circ \omega_{\boldsymbol{\lambda}}^q(T', T) = \Omega_A^q(\mathsf{B}(T'), \mathsf{B}(T)) \circ \rho^q$$

   and hence $\Omega_A^q(\mathsf{B}(T'), \mathsf{B}(T)) = \Omega_{\mathcal{C}}^q(T', T)$.
2. *Let $\varphi^q : K^q \twoheadrightarrow H^q(\mathsf{M}, \mathcal{L})$ be the natural projection. Then there is an isomorphism $\tau^q : A^q \to K^q$ so that a Gauss-Manin connection endomorphism $\Omega_{\mathcal{L}}^q(\mathsf{B}(T'), \mathsf{B}(T))$ in local system cohomology is determined by the equation*

$$\varphi^q \circ \tau^q \circ \omega^q_\lambda(T', T) = \Omega^q_{\mathcal{L}}(\mathsf{B}(T'), \mathsf{B}(T)) \circ \varphi^q \circ \tau^q.$$

Proof. Assume that the weights are T-nonresonant. Then all relevant information is determined by the combinatorial type, so we may write $H^q(T; \mathcal{L}) = H^q(\mathsf{M_b}; \mathcal{L_b})$ and use similar notation for $\mathcal{G}$. Recall from Theorems 2.2.5 and 2.2.6 that $H^q(T; \mathcal{L}) = 0$ for all $q \neq \ell$ and that $H^\ell(T; \mathcal{L}) \simeq H^\ell(A^\bullet(\mathcal{A}), a_\lambda)$.

We show first that a Gauss-Manin connection matrix $\Omega^\ell_{\mathcal{L}}(\mathsf{B}(T'), \mathsf{B}(T)) = \Omega^\ell_A(\mathsf{B}(T'), \mathsf{B}(T))$ in the $\beta\mathbf{nbc}$ basis for the unique nonvanishing local system cohomology group is induced by an endomorphism $\tilde{\Omega}_{\mathcal{L}}$ of the top cohomology of the complement of a general position arrangement.

Let γ be a simple loop in $\mathsf{B}(T)$ linking $\mathsf{B}(T')$ in the closure. Let Γ be a compatible class in the sense of Theorem 2.4.2. It induces an automorphism $\Psi_{\mathcal{G}}(\Gamma)$ of $H^\ell(\mathcal{G}; \mathcal{L})$. Choose a connection matrix $\tilde{\Omega}_{\mathcal{L}}$ so that $\Psi_{\mathcal{G}}(\Gamma) = \exp(-2\pi \mathrm{i}\, \tilde{\Omega}_{\mathcal{L}})$. Since the map $\bar{s}$ is surjective, the endomorphism $\tilde{\Omega}_{\mathcal{L}} : H^\ell(\mathcal{G}) \to H^\ell(\mathcal{G})$ induces an endomorphism $H^\ell(T) \to H^\ell(T)$, which we denote by $\Omega_{\mathcal{L}}$. We assert that $\Omega_{\mathcal{L}}$ is a connection matrix for $\Psi_T(\gamma)$. It suffices to show that $\exp(-2\pi \mathrm{i}\, \Omega_{\mathcal{L}})$ is conjugate to $\Psi_T(\gamma)$. By construction we have $\bar{s} \circ \tilde{\Omega}_{\mathcal{L}} = \Omega_{\mathcal{L}} \circ \bar{s}$. From this it follows that $\bar{s} \circ \exp(-2\pi \mathrm{i}\, \tilde{\Omega}_{\mathcal{L}}) = \exp(-2\pi \mathrm{i}\, \Omega_{\mathcal{L}}) \circ \bar{s}$. By Theorem 2.4.2 and Theorem 2.5.1, up to conjugacy, we have $\bar{s} \circ \Psi_{\mathcal{G}}(\Gamma) = \Psi_T(\gamma) \circ \bar{s}$. Hence, $\exp(-2\pi \mathrm{i}\, \Omega_{\mathcal{L}})$ is conjugate to $\Psi_T(\gamma)$.

$$
\begin{array}{ccc}
H^\ell(\mathcal{G}; \mathcal{L}) & \xrightarrow{\ \bar{s}\ } & H^\ell(T; \mathcal{L}) \\
\Big\downarrow{\tilde{\Omega}_{\mathcal{L}}} & & \Big\downarrow{\Omega_{\mathcal{L}}} \\
H^\ell(\mathcal{G}; \mathcal{L}) & \xrightarrow{\ \bar{s}\ } & H^\ell(T; \mathcal{L})
\end{array}
\tag{2.5}
$$

Next we show that this connection is a specialization of a formal connection. Aomoto and Kita [2] computed $\tilde{\Omega}_{\mathcal{L}}(\mathsf{B}(\mathcal{G}_J), \mathsf{B}(\mathcal{G}))$ where $\mathcal{G}_J$ is the combinatorial type whose only dependent set is the $(\ell + 1)$-tuple J. The formulas for $\tilde{\omega}_J$ were chosen to imply that $\Omega^\ell_{\mathcal{L}}(\mathsf{B}(\mathcal{G}_J), \mathsf{B}(\mathcal{G})) = \Omega^\ell_{\mathcal{C}}(\mathcal{G}_J, \mathcal{G})$. Theorem 2.4.1 and Theorem 2.4.2 imply that $\tilde{\Omega}_{\mathcal{C}}$ equals $\tilde{\Omega}_{\mathcal{C}}(T', T)$ up to conjugacy. Since $\tilde{\Omega}_{\mathcal{C}}(T', T)$ is induced by the specialization $\tilde{\omega}^\ell_\lambda(T', T)$ of the endomorphism $\tilde{\omega}^\ell(T', T)$ of the Aomoto complex of a general position arrangement, it follows from Theorem 1.10.1 that $\Omega^\ell_{\mathcal{L}}(\mathsf{B}(T'), \mathsf{B}(T))$ is induced by the specialization $\omega^\ell_\lambda(T', T)$ of the endomorphism $\omega^\ell(T', T)$ of the Aomoto complex of type T. The same holds for $q < \ell$ vacuously.

Thus, for T-nonresonant weights, the endomorphism $\omega(T', T)$ induces both the local system Gauss-Manin connection and the combinatorial Gauss-Manin connection in the $\beta\mathbf{nbc}$ basis for all q. Since $\omega(T', T)$ is a holomorphic map in the variables $\mathbf{y}$ and the set of T-nonresonant weights is open and dense, $\omega(T', T)$ induces a Gauss-Manin connection endomorphism for all weights in either cohomology theory. $\qquad\square$

Note that $\Omega^q_{\mathcal{L}}(\mathsf{B}(T'), \mathsf{B}(T)) = \Omega^q_{\mathcal{C}}(T', T)$ not only for T-nonresonant weights but also for resonant weights which are T-combinatorial. On the

other hand, there are examples of resonant weights which are not combi-
natorial where $H^q(\mathsf{M}; \mathcal{L}) \neq H^q(A^\bullet(\mathcal{A}), a_\lambda)$, see [48]. In these cases the groups
$H^q(\mathsf{M}; \mathcal{L})$ are not known to be combinatorial invariants. The point is that the
combinatorialcohomology bundle is locally trivial over the entire moduli space
of type $\mathcal{T}$, while the local system cohomology bundle is known to be locally
trivial only over connected components of the moduli space.

Nevertheless, the second part of Theorem 2.5.3 asserts that the associated
Gauss-Manin connection is induced by the combinatorially defined endomor-
phism $\omega(\mathcal{T}', \mathcal{T})$ of the Aomoto complex in all cases. The maps φ^q and τ^q
contain the additional (generally unknown) information.

Added in Proof: The eigenvalues of these Gauss-Manin connections have
been obtained for nonresonant weights in [17] and for all weights in [18].

Example

In order to find a substantial application of the second part of Theorem 2.5.3
we would have to find a type $\mathcal{T}$ which admits resonant weights that are not
combinatorial. All known examples are fairly large, see Exercise 4 in Section
2.6. We avoid such a large calculation by doing the next best thing. In what
follows we describe all the necessary steps for the application of Theorem
2.5.3 in the example of four lines in Figure 1.8 of Section 1.11. Here all resonant weights are combinatorial, and the reader is encouraged to find the
corresponding connection matrices by combinatorial methods.

The key difficulty is that we must compute the groups $H^q(\mathsf{M}, \mathcal{L})$ directly.
The universal complex $\big(\mathsf{K}^\bullet, \Delta^\bullet(\mathbf{x})\big)$ and Aomoto complex $\big(\mathsf{A}^\bullet, a_\mathbf{y}\big)$ of $\mathcal{A}$ are
recorded in [14, §6.1]. The universal complex is equivalent to the cochain
complex of the maximal abelian cover of the complement M. For any $\mathbf{t} \in (\mathbb{C}^*)^4$,
the specializations at $\mathbf{t}$ of the two complexes are quasi-isomorphic. The latter
complex may be obtained by applying the Fox calculus to a presentation of
the fundamental group of the complement, see for instance [19].

A presentation of the fundamental group of the complement is

$$\pi_1(M) = \langle x_1, x_2, x_3, x_4 \mid [x_3 x_1, x_2], [x_1 x_2, x_3], [x_i, x_4] \text{ for } i = 1, 2, 3 \rangle \quad (2.6)$$

where $[\alpha, \beta] = \alpha\beta\alpha^{-1}\beta^{-1}$. Check that the set of relators $[x_1, x_2, x_3]$ introduced
in Section 2.1 may be written as the first two commutators. The coboundary
map $\Delta^1(\mathbf{x}) : \mathsf{K}^1 \to \mathsf{K}^2$ is obtained by taking the Fox derivatives of the relators.
These derivatives are defined as follows.

Let $\mathbb{F}$ be a free group on the generators $x_1, \ldots, x_n$, $\mathbb{ZF}$ its integral group
ring, and $\epsilon : \mathbb{ZF} \to \mathbb{Z}$ the augmentation map. To each x_i corresponds a *Fox
derivative* $\partial_i : \mathbb{ZF} \to \mathbb{ZF}$ defined by $\partial_i(1) = 0$, $\partial_i(x_j) = \delta_{ij}$, and $\partial_i(uv) = \partial_i(u)\epsilon(v) + u\partial_i(v)$.

We get the following matrix with rows indexed by the variables and
columns indexed by the relators:

$$\Delta^1 = \Delta^1(\mathbf{x}) = \begin{bmatrix} x_3 - x_2 x_3 & 1 - x_3 & 1 - x_4 & 0 & 0 \\ x_1 x_3 - 1 & x_1 - x_1 x_3 & 0 & 1 - x_4 & 0 \\ 1 - x_2 & x_1 x_2 - 1 & 0 & 0 & 1 - x_4 \\ 0 & 0 & x_1 - 1 & x_2 - 1 & x_3 - 1 \end{bmatrix}.$$

The coboundary map in the Aomoto complex $a_{\mathbf{y}} : A^1 \to A^2$ has matrix

$$\mu^1 = \mu^1(\mathbf{y}) = \begin{bmatrix} -y_2 & -y_3 & -y_4 & 0 & 0 \\ y_{13} & -y_3 & 0 & -y_4 & 0 \\ -y_2 & y_{12} & 0 & 0 & -y_4 \\ 0 & 0 & y_1 & y_2 & y_3 \end{bmatrix}.$$

Its rows correspond to the products $(y_1 a_1 + y_2 a_2 + y_3 a_3 + y_4 a_4) a_i$ for $i = 1, 2, 3, 4$. Its columns are indexed by the ordered **nbc** basis for $A^2(\mathcal{A})$: $a_{12}, a_{13}, a_{14}, a_{24}, a_{34}$. Recall that $a_{23} = -a_{12} + a_{13}$. Note that μ^1 equals Δ^1 linearized: replace x_i by $1 + y_i$ and retain only the linear terms. A nontrivial resonant local system $\mathcal{L}$ on M corresponds to a point $\mathbf{1} \neq \mathbf{t} \in (\mathbb{C}^*)^4$ satisfying $t_1 t_2 t_3 = 1$ and $t_4 = 1$. For each such $\mathbf{t}$ we have $H^2(\mathsf{M}; \mathcal{L}) \simeq \mathbb{C}^3$. Define $\Xi : \Lambda^5 \to \Lambda^3$ and $\Upsilon : R^5 \to R^3$ by

$$\Xi = \begin{bmatrix} x_1 x_2 - 1 & 0 & x_1 x_2 + x_4 - 2 \\ x_2 - 1 & 0 & x_2 - 1 \\ 0 & x_2 - 1 & 0 \\ 0 & 1 - x_1 & x_3 - 1 \\ 0 & 0 & 1 - x_2 \end{bmatrix} \quad \text{and} \quad \Upsilon = \begin{bmatrix} y_{12} & 0 & y_{124} \\ y_2 & 0 & y_2 \\ 0 & y_2 & 0 \\ 0 & -y_1 & y_3 \\ 0 & 0 & -y_2 \end{bmatrix}.$$

Note that Υ is the linearization of Ξ. Since $\mathcal{L}$ is nontrivial, $t_i \neq 1$ for some i. Assume, without loss, that $t_2 \neq 1$. For each $\mathbf{t}$ satisfying $t_1 t_2 t_3 = 1$, $t_4 = 1$, and this condition, check that $\operatorname{rank} \Xi(\mathbf{t}) = 3$ and $\Xi(\mathbf{t}) \circ \Delta^1(\mathbf{t}) = 0$. So the projection $\mathbb{C}^5 \simeq K^2 \twoheadrightarrow H^2(\mathsf{M}; \mathcal{L}) \simeq \mathbb{C}^3$ may be realized as the evaluation $\varphi^2 = \Xi(\mathbf{t})$.

Via the map Υ, the endomorphisms $\omega(\mathcal{T}_j, \mathcal{T})$ induce maps $\Omega_j : R^3 \to R^3$, which satisfy $\Upsilon \circ \omega^2(\mathcal{T}_j, \mathcal{T}) = \Omega_j \circ \Upsilon$. These maps have matrices

$$\Omega_1 = \begin{bmatrix} 0 & 0 & y_{124} \\ 0 & 0 & -y_3 \\ 0 & 0 & -y_{12} \end{bmatrix}, \quad \Omega_2 = \begin{bmatrix} y_{12} & 0 & y_{124} \\ 0 & y_{12} & -y_3 \\ 0 & 0 & 0 \end{bmatrix}, \quad \Omega_3 = \begin{bmatrix} y_4 & -y_4 & -y_{1234} \\ -y_{123} & y_{123} & -y_{1234} \\ 0 & 0 & y_{1234} \end{bmatrix}.$$

Let $\boldsymbol{\lambda}$ be a collection of weights corresponding to $\mathbf{t}$. Note that $\lambda_2 \notin \mathbb{Z}$ since $t_2 \neq 1$. Hence, $\Upsilon(\boldsymbol{\lambda}) : \mathbb{C}^5 \twoheadrightarrow \mathbb{C}^3$ is surjective for all such $\boldsymbol{\lambda}$. Identify $H^2(\mathsf{M}; \mathcal{L}) = \mathbb{C}^3$, and let $\tau^2 : A^2 \to K^2$ be an isomorphism for which $\varphi^2 \circ \tau^2 = \Upsilon(\boldsymbol{\lambda})$. Theorem 2.5.3 implies that a Gauss-Manin connection matrix $\Omega_{\mathcal{L}}^2(\mathsf{B}(\mathcal{T}_j), \mathsf{B}(\mathcal{T}))$ in $H^2(\mathsf{M}; \mathcal{L})$ corresponding to the degeneration $\mathcal{T}_j$ of $\mathcal{T}$ satisfies $\Upsilon(\boldsymbol{\lambda}) \cdot \omega_{\boldsymbol{\lambda}}^q(\mathcal{T}', \mathcal{T}) = \Omega_{\mathcal{L}}^q(\mathsf{B}(\mathcal{T}_j), \mathsf{B}(\mathcal{T})) \cdot \Upsilon(\boldsymbol{\lambda})$. Since the equality $\Upsilon \circ \omega^2(\mathcal{T}_j, \mathcal{T}) = \Omega_j \circ \Upsilon$ holds in the Aomoto complex, the specialization $\mathbf{y} \mapsto \boldsymbol{\lambda}$ yields connection matrices $\Omega_{\mathcal{L}}^2(\mathsf{B}(\mathcal{T}_j), \mathsf{B}(\mathcal{T})) = \Omega_j(\boldsymbol{\lambda})$.

The endomorphisms $\Omega_{\mathcal{L}}^1(\mathsf{B}(\mathcal{T}_j), \mathsf{B}(\mathcal{T}))$ may be determined by similar methods. As noted in [14], the endomorphism $\Omega_{\mathcal{L}}^1(\mathsf{B}(\mathcal{T}_2), \mathsf{B}(\mathcal{T}))$ corresponds to the automorphism of $H^1(M; \mathcal{L}) \simeq \mathbb{C}$ given by multiplication by $t_1 t_2$, so it has matrix $[\lambda_{12}]$. Note that $\lambda_{12} \in \mathbb{Z}$ if $t_1 t_2 = 1$. The endomorphisms $\Omega^1_{\mathcal{L}}(\mathrm{B}(\mathcal{T}_1), \mathrm{B}(\mathcal{T}))$ and $\Omega^1_{\mathcal{L}}(\mathrm{B}(\mathcal{T}_3), \mathrm{B}(\mathcal{T}))$ are trivial.

Connection matrices corresponding to other codimension-one degenerations of $\mathcal{T}$ may be obtained by analogous calculations. Note however that the projection $\Upsilon(\lambda) : A^2 \rightarrow H^2(M; \mathcal{L})$ given above need not be relevant for all degenerations.

2.6 Exercises

1) Find a saturated chain of degenerations from the general position arrangement of 5 lines to the Selberg arrangement. (Saturated means that if $\mathcal{T}$ and $\mathcal{T}'$ are in the chain, then there is no realizable type $\mathcal{T}''$ with $\mathcal{T} > \mathcal{T}'' > \mathcal{T}'$.)

2) Find a second saturated chain for Problem 17. Do the two chains have the same length?

3) Consider the Selberg arrangement with $\lambda = 1/3(1, 1, 1, 1, -2)$. Show that λ is combinatorial. Compute $H^1(A, a_\lambda)$ to show that λ is resonant.

4) Define the Ceva(3) arrangement in $\mathbb{CP}^2$ by

$$Q = (u_1^3 - u_2^3)(u_1^3 - u_3^3)(u_2^3 - u_3^3).$$

Show that $\lambda = 1/3(1, 1, 1, 1, 1, 1, -2, -2, -2)$ is not $\mathcal{T}$-combinatorial. [Hint: Show that there are no integers $k_1, \ldots, k_9$ with $\sum_{i=1}^{9} k_i = 0$ so that the weights $\lambda'_i = \lambda_i + k_i$ are $\mathcal{T}$-combinatorial.]

References

1. Aomoto, K.: Hypergeometric functions, the past, today, and ... (from the complex analytic point of view), Sugaku Expositions, **9** (1996), 99–116
2. Aomoto, K., Kita, M.: *Hypergeometric functions* (in Japanese), Springer-Verlag, 1994
3. Arapura, D.: Geometry of cohomology support loci for local systems I, J. Alg. Geom. **6** (1997), 563–597
4. Artal Bartolo, E., Carmona Ruber, J., Cogolludo Agustin, J.I., Marco Buzunáriz, M.A.: Invariants of combinatorial line arrangements and Rybnikov's example, preprint, 2004
5. Arvola, W.: The fundamental group of the complement of an arrangment of complex hyperplanes, Topology **31** (1992), 757–765
6. Bidigaire, T.P., Hanlon, P., Rockmore, D.: A combinatorial description of the spectrum for the Tsetlin library and its generalization to hyperplane arrangements, Duke Math. J. **99** (1999), 135–174
7. Billera, L.J., Brown, K.S., Diaconis, P.: Random walks and plane arrangements in three dimensions, Amer. Math. Monthly, **106** (1999), 502–524
8. Brieskorn, E.: Sur les groupes de tresses. In: Séminaire Bourbaki 1971/72. Lecture Notes in Math. **317**, Springer Verlag, 1973, pp. 21–44
9. Brown, K.S., Diaconis, P.: Random walks and hyperplane arrangements, Ann. Probab., **26** (1998), 1813–1854
10. Brylawski, T. and Varchenko, A.: The determinant formula for a matroid bilinear form, Adv. Math. **129** (1997), 1–24
11. Cohen, D.: Cohomology and intersection cohomology of complex hyperplane arrangements, Adv. Math. **97** (1993), 231–266
12. Cohen, D.: On the cohomology of discriminantal arrangements and Orlik-Solomon algebras, in: Singularities and Arrangements, Sapporo-Tokyo 1998, Adv. Stud. Pure Math., **27**, Math. Soc. Japan, Tokyo 2000, 27–49
13. Cohen, D., Orlik, P.: Arrangements and local systems, Math. Res. Lett. **7** (2000), 299–316.
14. Cohen, D., Orlik, P.: Gauss-Manin connections for arrangements I, Eigenvalues, Compositio Math. **136** (2003), 299–316
15. Cohen, D., Orlik, P.: Gauss-Manin connections for arrangements II, Nonresonant weights, Amer. J. Math., **127** (2005), 569–594

16. Cohen, D., Orlik, P.: Gauss-Manin connections for arrangements III, Formal connections, Trans. Amer. Math. Soc., **357** (2005), 3031–3050

17. Cohen, D., Orlik, P.: Gauss-Manin connections for arrangements IV, Nonresonant eigenvalues, Comm. Math. Helv., to appear

18. Cohen, D., Orlik, P.: Stratified Morse Theory in Arrangements, Pure and Appl. Math. Quarterly, **2** (2006), 673–697

19. Cohen, D., Suciu, A.: On Milnor fibrations of arrangements, J. London Math. Soc. **51** (1995), 105–119

20. Cohen, D., Suciu, A.: Characteristic varieties of arrangements, Math. Proc. Cambridge Philos. Soc. **127** (1999), 33–54

21. Deligne, P.: Equations Différentielles à Points Singuliers Réguliers, Lecture Notes in Mathematics, **163** Springer Verlag, 1970

22. Deligne, P.: Les immeubles des groupes de tresses généralisés, Invent. math. **17** (1972), 273–302

23. Dimca, A., Papadima, S.: Hypersurface complements, Milnor fibers, and higher homotopy groups of arrangements, Ann. Math. **158** (2003), 473–507

24. Esnault,H., Schechtman,V., Viehweg, V.: Cohomology of local systems on the complement of hyperplanes, Invent. Math. **109** (1992), 557–561. Erratum, ibid. **112** (1993), 447

25. Falk, M.: Arrangements and cohomology, Ann. Comb. **1** (1997), 135–157

26. Falk, M., Randell, R.: The lower central series of a fiber-type arrangement, Invent. math. **82** (1985), 77–88

27. Falk, M., Randell, R.: On the homotopy type of arrangements, *Complex Analytic Singularities*, Adv. Studies in Math. **8**, North Holland, 1986, 101–124

28. Falk, M., Randell, R.: On the homotopy type of arrangements, II *Arrangements-Tokyo 1998*, Adv. Studies in Math. **27**, Math. Soc. Japan, 2000, 93–125

29. Falk, M., Terao, H.: βNBC-bases for cohomology of local systems on hyperplane complements, Trans. AMS, **349** (1997), 189–202

30. Gelfand, I.M.: General theory of hypergeometric functions, Soviet Math. Dokl. **33** (1986), 573–577

31. Kamiya, H., Orlik, P., Takemura, A., Terao, H.: Arrangements and Ranking Patterns, Ann. Combinatorics, **10** (2006), 267–284

32. Kemeny, J.G., Snell, J.L.: Finite Markov chains, Van Nostrand, Princeton, NJ 1960

33. Kobayashi, S.: *Differential geometry of complex vector bundles*, Princeton Univ. Press, Princeton, NJ, 1987.

34. Libgober, A.: Characteristic varieties of algebraic curves, in *Applications of algebraic geometry to coding theory, physics and computation* (Eilat, 2001) 215–254, NATO Sci. Ser. II Math. Phys. Chem. **36**, Kluwer Publ., 2001

35. Libgober, A.: First order deformations for rank one local systems with non vanishing cohomology, Top. Appl. **118** (2002), 159–168

36. Libgober,A., Yuzvinsky, S.: Cohomology of the Orlik-Solomon algebras and local systems, Compositio Math. **121** (2000), 337–361

37. Orlik, P., Solomon, L.: Combinatorics and topology of complements of hyperplanes, Invent. math. **56** (1980), 167–189

38. Orlik, P., Terao, H.: *Arrangements of Hyperplanes*, Springer Verlag, 1992

39. Orlik, P., Terao, H.: Arrangements and Hypergeometric Integrals, Math. Soc. Japan Memoirs **Vol. 9**, 2001

40. Randell, R.: The fundamental group of the complement of a union of complex hyperplanes, Invent. math. **69** (1982) 103–108. Correction: Invent. math. **80** (1985) 467–468

41. Randell, R.: Lattice-isotopic arrangements are topologically isomorphic, Proc. Amer. Math. Soc. **107** (1989), 555–559

42. Randell, R.: Morse theory, Milnor fibers and minimality of hyperplane arrangements, Proc. Amer. Math. Soc. **130** (2002), 2737–2743

43. Rybnikov, G.: On the fundamental group of a complex hyperplane arrangement, DIMACS Technical Reports, 13, 1994

44. Schechtman, V., Terao, H., Varchenko, A.: Cohomology of local systems and the Kac-Kazhdan condition for singular vectors, J. Pure Appl. Algebra **100** (1995), 93–102

45. Schechtman, V., Varchenko, A.: Hypergeometric solutions of Knizhnik-Zamolodchikov equations, Lett. Math. Phys. **20** (1990), 279–283

46. Schechtman, V., Varchenko, A.: Arrangements of hyperplanes and Lie algebra homology, Invent. Math. **106** (1991), 139–194

47. Selberg, A.: Bemerkninger om et multipelt integral, Norsk matematisk tidsskrift **26** (1944) 71–78, In: Collected Papers **1**, Springer Verlag, 1989, 204–211

48. Suciu, A. I.: Translated tori in the characteristic varieties of complex hyperplane arrangements, Top. Appl. **118** (2002), 209–223

49. Terao, H.: Modular elements of lattices and topological fibrations, Adv. Math. **62** (1986), 135–154

50. Terao, H.: Moduli space of combinatorially equivalent arrangements of hyperplanes and logarithmic Gauss-Manin connections, Topology Appl. **118** (2002), 255–274.

51. Varchenko, A.: *Multidimensional hypergeometric functions and representation theory of Lie groups*, Adv. Studies in Math. Physics **21** World Scientific Publishers 1995

52. Varchenko, A.: *Special functions, KZ type equations, and representation theory*, CBMS Lecture Notes **98** Amer. Math. Soc., 2003

53. Yuzvinsky, S.: Orlik-Solomon algebras in algebra and topology, Russian Math. Surveys **56** (2001), 293–364

54. Zaslavsky, T.: *Facing up to arrangements: Face-count formulas for partitions of space by hyperplanes*, Memoirs Amer. Math. Soc. **154** 1975

55. Ziegler, G.: Matroid shellability, β–systems, and affine arrangements, J. Alg. Combinatorics **1** (1992), 283–300

Part II

Discrete Morse Theory
and Free Resolutions

Volkmar Welker

1

Introduction

1.1 Overview

These lecture notes present topics from Algebraic Combinatorics that lie on the borderline to Algebraic Topology and Commutative Algebra. In particular, we will present and review combinatorial and geometric methods for studying minimal free resolutions of ideals in polynomial rings. Before we give the exposition of the topics we will spend time in order to outline the basic mathematical theory behind. Thus these notes will also include definitions and examples for CW-complexes and free resolutions. All this basic material is geared towards the applications given in the later sections and is therefore not presented in utmost generality. A comprehensive exposition of the interaction between Combinatorics and Commutative Algebra and the history of this interaction can be found in the books by Miller and Sturmfels [35] and Stanley [57].

Our methods will be algebraic and topological. The main idea, first employed in [6], is that in many cases one can associate to a free resolution of an ideal a topological space – a CW-complex – that carries in its cellular structure the algebraic structure of the free resolution. It will be demonstrated that methods from Topological Combinatorics – Discrete Morse Theory – can be used to minimize the given resolution. This idea was first developed in [3] and more examples and applications beyond the ones given in these notes can also be found in [4]. Recently, three independent generalizations to arbitrary algebraic complexes [32], [51] and [27] have been successfully applied. But, we will stick to the setting of cellular resolutions and Discrete Morse Theory applied to them. We do so, since we think that the additional structure of a geometric object sheds further insight into the algebraic object of a minimal free resolution and also poses several intriguing questions about the geometry of cell complexes (see e.g. [60]).

From the point of view of Combinatorics minimal free resolutions can serve as a tool for the study of invariants of simplicial complexes. Since the actual relation to Combinatorics will be less obvious in the main body of these notes, we devote a major part of this introduction to establishing the link between the topics touched in the notes and their applications in Combinatorics.

On the other hand combinatorial methods, among them the ones presented in these lecture notes, have been successfully applied to obtain results in Commutative Algebra and Combinatorial Topology. Combinatorial methods have led to explicit constructions of minimal free resolutions of ideals that arise in Commutative Algebra (see e.g. [6]) and combinatorial methods have led to the calculation of explicit homotopy types of spaces and (co)-homology groups of spaces that arise in Algebraic Topology (see e.g. [64]). Unfortunately these applications are either technical or require techniques that are beyond the scope of this manuscript. Therefore we do not provide an exposition here and refer the reader to the book [35]. For more basic facts from Topological Combinatorics we point to the article by Björner [9] and for more information on Discrete Morse Theory to the article by Forman [20]. The very nice book by Hibi [26] is another source for additional information.

Fig. 1.1. shows the main objects of these notes positioned in between the three major mathematical fields and their relevant subfields or relevant objects.

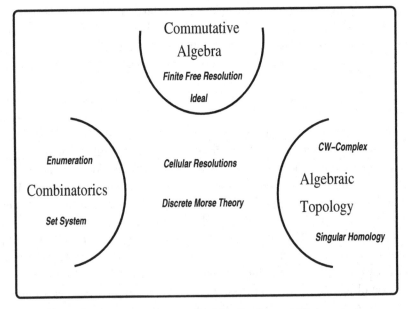

Fig. 1.1. Topics touched by the Lecture Notes

The notes are divided into four chapters. The first chapter contains the introduction with an overview of the notes and sections motivating the study of minimal free resolutions and cellular resolutions in Algebraic Combinatorics. In the second chapter we give an exposition of the basic mathematical objects studied. In particular, we define multigraded minimal free resolutions, CW-complexes and cellular resolutions. The third chapter focuses on cellular resolutions and provides a long list of examples. Then in the fourth chapter we introduce Discrete Morse Theory as developed by Forman [20] and show how it can be applied to minimize cellular resolutions. Each chapter is supplemented by a section containing open problems and exercises.

These notes grew out of the lectures given at the 2003 summer school on 'Algebraic Combinatorics' in Nordfjordeid. I would like to thank the organizers of this event for their work which created a very productive and pleasant atmosphere. I very much enjoyed the week teaching in this wonderful place together with Sergey Fomin and Peter Orlik. Those parts of my lecture which are not 'standard' Algebraic Combinatorics – the application of discrete Morse theory to free resolutions – were developed in joint work with Ekki Batzies in [3] and [4]. While developing these ideas in [3] Ekki found an alternative approach to the standard discrete Morse theory. Our approach to discrete Morse theory in these notes is based on his ideas. I am grateful to Ekki for letting me include his ideas in this notes. I also thank Anna Thursby for many helpful conversations while working on these notes.

Let us now turn to some applications of (Commutative) Algebra in Combinatorics which are supposed to motivate the subsequent chapters and put them in perspective.

1.2 Enumerative and Algebraic Invariants of Simplicial Complexes

Recall that an *(abstract) simplicial complex* Δ on the (finite) ground set Ω is a subset $\Delta \subseteq 2^\Omega$ of the power set of Ω such that for any $A \subseteq B \in \Delta$ it follows that $A \in \Delta$. We will often assume that $\Omega = [d] := \{1, \ldots, d\}$ is the set of the first d integers. Even though the concept of a simplicial complex appears to be simple minded, it indeed captures many objects that are of central interest in Algebraic Combinatorics.

Example 1.2.1. A matroid M on the ground set Ω is a set of subsets of Ω such that for any $A \subseteq B \in M$ we have $A \in M$ – which implies that M is a simplicial complex – and such that if $A, B \in M$ and $|A| = |B| + 1$ then there is a $y \in A \setminus B$ such that $B \cup \{y\} \in M$. The collection of sets in M is called the set of independent sets of M. There are many equivalent definitions of a matroid (see [61]) but for the point of view we take in these notes this definition is the most suitable. Let us give a few concrete examples of matroids.

1. $U_{d,l} = \{A \subseteq [d] \mid \#A \leq l\}$ is the complex of all subsets of $[d]$ of cardinality $\leq l$. It is easily seen that $U_{d,l}$ is a matroid and it is called uniform matroid.
2. Let $\Omega \subseteq V$ be a finite subset of a vectorspace V over a field k and set $\mathfrak{Lin}_\Omega$ to be the set of subsets of Ω that are linearly independent over k. Then basic linear algebra shows that $\mathfrak{Lin}_\Omega$ is a matroid. Thus matroids can be seen as a generalization of the concept of a vectorspace.
3. Let $\Omega = \mathcal{A}$ be a finite set of hyperplanes in a vectorspace V over a field k. Let $\mathfrak{A}_\mathcal{A}$ be the set of subsets B of $\mathcal{A}$ such that $\mathrm{codim}_k \bigcap_{H \in B} H = \#B$ – the intersection over the empty set being the ambient vectorspace V. Then again $\mathfrak{A}_\mathcal{A}$ is a matroid, which coincides with the matroid $\mathfrak{Lin}_\Omega$ for Ω a set of linear forms in the dual space V^* defining the hyperplanes in $\mathcal{A}$.

At this point Algebraic Combinatorics and Enumerative Combinatorics overlap. Before we are going to demonstrate this connection we will have to make a few more definitions.

For a simplicial complex Δ and $A \in \Delta$ the dimension $\dim(A)$ is given by $\#A - 1$. The dimension $\dim \Delta$ of the simplicial complex Δ is $\max_{A \in \Delta} \dim(A)$.

Example 1.2.2. The rank $\mathrm{rk}(M)$ of a matroid is the maximal cardinality of its independent sets. Thus we have $\dim M = \mathrm{rk}M - 1$. In particular, $\dim U_{d,l} = l - 1$, $\dim \mathfrak{Lin}_\Omega = \dim \mathrm{span}_k(\Omega) - 1$ and $\dim \mathfrak{A}_\mathcal{A} = \mathrm{codim} \bigcap_{H \in \mathcal{A}} H - 1$.

We call an element $A \in \Delta$ of a simplicial complex Δ a face of Δ. If the face A is maximal in Δ with respect to inclusion (i.e., $A \subseteq B \in \Delta$ implies $B = A$) then we call A a facet. If $\dim(A) = \dim(\Delta)$ for all facets A of Δ then we call Δ a pure simplicial complex .

Example 1.2.3. A matroid M is a pure simplicial complex. For if $|A| < |B|$ for two faces A and B of M then there is a face $C \subseteq B$ such that $|C| = |A| + 1$ and hence $A \cup \{y\}$ is in M for some $y \in C \setminus A$. Thus all facets have the same dimension which then must be $\dim(\Delta)$.

The f-vector of a simplicial complex Δ is the vector

$$f(\Delta) = (f_{-1}, \ldots, f_{\dim(\Delta)}),$$

where f_i is the number of i-dimensional faces of Δ.

Example 1.2.4. For the uniform matroid $U_{d,l}$ the dimension is given by $\dim U_{d,l} = l - 1$ and the $f(U_{d,l}) = (\binom{d}{0}, \binom{d}{1}, \ldots, \binom{d}{l})$.

A first question to be asked is: When is a given vector the f-vector of a simplicial complex ? The answer to this question is provided by the famous Kruskal-Katona Theorem (see for example [1]).

For the formulation of the theorem the following lemma introduces the binomial expansion of a natural number.

Lemma 1.2.5. *For natural numbers n and i there exists a unique set of natural numbers $j \le i$ and $a_i > \cdots > a_j \ge j \ge 1$ such that*

$$n = \binom{a_i}{i} + \cdots + \binom{a_j}{j}.$$

For natural numbers i and n we denote by $n^{(i)}$ the number $\binom{a_i}{i+1} + \cdots + \binom{a_j}{j+1}$, where $n = \binom{a_i}{i} + \cdots + \binom{a_j}{j}$ is the binomial expansion.

Theorem 1.2.6 (Kruskal-Katona Theorem). *A vector $(1, f_0, \ldots, f_{D-1})$ of nonnegative integers is the f-vector of a $(D-1)$-dimensional simplicial complex if and only if $f_{i+1} \le f_i^{(i+1)}$ for $0 \le i \le D-2$.*

Here we have a first chance to pass to algebra. For a simplicial complex Δ on $[d]$ let I_Δ be the ideal in the polynomial ring $k[x_1, \ldots, x_d]$ generated by monomials $\underline{x}_A = \prod_{i \in A} x_i$ for subsets $A \subseteq [d]$ such that $A \notin \Delta$. The ideal I_Δ is called the Stanley-Reisner ideal of Δ. A minimal generating system is given by the monomials $\underline{x}_A$ for the inclusionwise minimal non-faces A of Δ. Clearly, if $A \subseteq B$ and $A, B \notin \Delta$ then $\underline{x}_B$ is a multiple of $\underline{x}_A$ and therefore lies in any ideal with $\underline{x}_A$ in its generating set. By $k[\Delta]$ we denote the quotient $k[x_1, \ldots, x_d]/I_\Delta$ which is called Stanley-Reisner ring of Δ.

Example 1.2.7. If $\Delta = M$ is a matroid over Ω then a minimal non-face of M is also called circuit – or minimal dependent set. Thus I_M is minimally generated by the $\underline{x}_A$ for the circuits A of M. For $M = U_{d,l}$ and $A \subseteq [d]$ we have $A \notin U_{d,l}$ if and only if $\#A \ge l + 1$. Thus an $A \notin U_{d,l}$ is inclusionwise minimal if and only if $\#A = l + 1$. Hence $I_{U_{d,l}}$ is in this case minimally generated by the $\underline{x}_A$ for $\#A = l + 1$.

The concept of a Stanley-Reisner ring is connected to the enumerative invariant 'f-vector' by counting numbers of monomials. A monomial in $k[x_1, \ldots, x_d]$ is a product $\underline{x}^\alpha = x_1^{\alpha_1} \cdots x_d^{\alpha_d}$ for $\alpha = (\alpha_1, \ldots, \alpha_d) \in \mathbb{N}^d$. Let us count the number of monomials whose image under the canonical projection $\pi : k[x_1, \ldots, x_d] \to k[\Delta]$ is non-zero. For this to happen we must have that the monomial is not divisible by a monomial $\underline{\mathbf{x}}_A$ for a minimal non-face A of Δ. But this condition applied to a monomial $\underline{\mathbf{x}}^\alpha$ means that the support

$$B = \mathrm{supp}(\underline{\mathbf{x}}^\alpha) := \{i \mid \alpha_i > 0\}$$

must lie in Δ. Thus $\underline{\mathbf{x}}^\alpha$ is a multiple of $\underline{\mathbf{x}}_B$ inside the polynomial ring $k[x_i \mid i \in B]$. If we define the degree $\deg(\underline{\mathbf{x}}^\alpha)$ of $\underline{\mathbf{x}}^\alpha$ by $|\alpha| := \alpha_1 + \cdots + \alpha_d$ then this argumentation implies that there are $\binom{j-1}{\dim(B)}$ monomials of degree j and support B that are not in the kernel of π. This shows that the generating function $H_B(t) = \displaystyle\sum_{j \geq \dim(B)} \binom{j-1}{\dim(B)} t^{\dim(B)+1}$ of the monomials with support

B enumerated by degree equals $H_B(t) = \dfrac{t^{\dim(B)}}{(1-t)^{\dim(B)}}$. Hence the generating function $\mathrm{Hilb}_\Delta(t)$ of the monomials in $k[x_1,\ldots,x_d]$ that do not lie in the kernel of the canonical projection to $k[\Delta]$ by degree is given as $\mathrm{Hilb}_\Delta(t) = \displaystyle\sum_{B \in \Delta} H_B(t)$. Thus

$$\mathrm{Hilb}_\Delta(t) = \sum_{i=0}^{\dim(\Delta)+1} \frac{f_{i-1} t^i}{(1-t)^i}.$$

The generating function $\mathrm{Hilb}_\Delta(t)$ is also called the Hilbert-series of $k[\Delta]$. This fact is due to the observation that the images of the monomials of degree j that do not lie in the kernel of the canonical projection from $k[x_1,\ldots,x_d]$ to $k[\Delta]$ form a basis for the vectorspace of homogeneous polynomials of degree j in $k[\Delta]$. In general, for a graded k-algebra the Hilbert-series is the generating series of the dimensions of the graded pieces (see [12, Chapter 5], [18] for more on Hilbert-series of Stanley-Reisner rings and general graded algebras).

It is natural to write $\mathrm{Hilb}_\Delta(t)$ as a rational function with a common denominator. We then get:

$$\mathrm{Hilb}_\Delta(t) := \frac{\displaystyle\sum_{i=0}^{\dim(\Delta)+1} f_{i-1} t^i (1-t)^{\dim(\Delta)+1-i}}{(1-t)^{\dim(\Delta)+1}}.$$

If we expand the numerator polynomial in the t^i then:

$$\mathrm{Hilb}_\Delta(t) = \frac{\displaystyle\sum_{i=0}^{\dim(\Delta)+1} h_i t^i}{(1-t)^{\dim(\Delta)+1}}.$$

The vector $h(\Delta) = (h_0, \ldots, h_{\dim(\Delta)+1})$ is then called the h-vector of Δ.

Remark 1.2.8.

1. $h_0 + \cdots + h_d = f_{\dim(\Delta)}.$
2. $h_0 = 1.$
3.

$$h_{\dim(\Delta)+1} = \sum_{i=0}^{\dim(\Delta)+1} (-1)^{\dim(\Delta)+1-i} f_{i-1} = (-1)^{\dim(\Delta)} \sum_{i\geq 0}(-1)^{i-1} f_{i-1}.$$

Note, that $\sum_{i\geq 0}(-1)^{i-1} f_{i-1}$ is also known as the reduced Euler-characteristic of Δ (see [37]). It is clear that the h-vector arises from the f-vector by an invertible linear transformation. Thus both carry the same amount of information. At first sight the f-vector appears to be the more natural invariant but then as we will see in Section 1.3 the h-vector turns out to be the 'better' invariant in many situations.

Example 1.2.9. If $\Delta = U_{d,l}$ is the uniform matroid then

$$\mathrm{Hilb}_{U_{d,l}}(t) = \frac{\sum_{i=0}^{l} \binom{d}{i} t^i (1-t)^{l-i}}{(1-t)^l}.$$

If one expands the numerator special values of l in terms of t-powers then we get:

$$l = 0 \qquad\qquad \mathrm{Hilb}_{U_{d,0}}(t) = 1$$
$$l = 1 \qquad\qquad \mathrm{Hilb}_{U_{d,1}}(t) = \frac{1 + (d-1)t}{(1-t)^1}$$
$$l = 2 \qquad \mathrm{Hilb}_{U_{d,2}}(t) = \frac{1 + (d-2)t + (d-2)u + \binom{d-1}{2}t^2}{(1-t)^2}$$
$$l = d-1 \qquad\qquad \mathrm{Hilb}_{U_{d,d-1}}(t) = \frac{1 + t + \cdots + t^d}{(1-t)^{d-1}}$$
$$l = d \qquad\qquad \mathrm{Hilb}_{U_{d,d}}(t) = \frac{1}{(1-t)^d}$$

There is one further way to calculate the h-vector of a simplicial complex from algebra. Here we introduce the concept of a minimal free resolution . This concept will be crucial for these notes and a more thorough definition and introduction will be given in Chapter 2.

Namely, let $S = k[x_1, \ldots, x_d]$ and

$$\mathcal{F}: 0 \xrightarrow{\partial_{r+1}} F_r \xrightarrow{\partial_r} \cdots \xrightarrow{\partial_1} F_0.$$

be a minimal free resolutions of $k[\Delta]$ over S, that is

(FR1) F_i is a free S-module for $0 \leq i \leq r$ (i.e. $F_i \cong S^{\beta_i}$ for some $\beta_i \geq 0$ in case F_i is finitely generated).

(FR2) The differentials ∂_i are S-module homomorphisms for $1 \leq i \leq r+1$.

(FR3) The sequence is exact; i.e. $\mathrm{Ker}\partial_i = \mathrm{Im}\partial_{i+1}$, $1 \leq i \leq r$.

(FR4) $\mathrm{CoKer}\partial_1 \cong k[\Delta]$.

(Min) $\mathrm{Im}\partial_i \subseteq \mathfrak{m}F_{i-1}$ for $1 \leq i \leq r+1$ and $\mathfrak{m} = \langle x_1, \ldots, x_d \rangle$ the unique graded maximal ideal in S.

An object satisfying conditions (FR1)-(FR4) is called free resolution of $k[\Delta]$ and (Min) is the minimality condition. If we choose bases for the F_i and express the ∂_i as matrices then (Min) is easily seen to be equivalent to

(Min') The matrices representing ∂_i, $1 \leq i \leq r+1$, contain no entry from $k \setminus \{0\}$.

It is well known (see for example [18, Chapter 20]) that minimal free resolutions always exist. Indeed more general, minimal free resolutions are studied for modules over rings. But here we confine ourselves to the combinatorial setting of the S-module $k[\Delta]$. It is easily seen that in a minimal free resolution of $k[\Delta]$ the F_i are always finitely generated. The sequence $(\beta_0, \ldots, \beta_r)$ is called the sequence of Betti numbers of $k[\Delta]$ – which can be shown [18, Chapter 20] to be independent of the chosen minimal free resolution. The number r is also called projective dimension of $k[\Delta]$. The Auslander-Buchsbaum theorem (see [18]) states $d - r = \mathrm{depth}(k[\Delta])$, where $\mathrm{depth}(k[\Delta])$ is the depth of $k[\Delta]$. For the moment we take this as the definition of depth. The depth will resurface and gain importance in Section 1.3, where we will give its usual definition. Actually we will see in examples that using the Auslander-Buchsbaum theorem we can sometimes calculate the depth from minimal free resolutions.

Now we want to put more structure on the free resolution. For this purpose we consider $S = k[x_1, \ldots, x_d]$ as an $\mathbb{N}^d$-graded or multigraded k-algebra; that is $S \cong \bigoplus_{\alpha \in \mathbb{N}^d} S_\alpha$, where S_α is the 1-dimensional k-vectorspace generated by $\mathbf{x}^\alpha$.

Since I_Δ is generated by monomials it follows as a k-vectorspace

$$I_\Delta \cong \bigoplus_{\substack{\alpha \in \mathbb{N}^d \\ \mathrm{supp}\, \mathbf{x}^\alpha \notin \Delta}} S_\alpha.$$

Thus the $\mathbb{N}^d$-grading carries over to the quotient $k[\Delta]$.

Now we require $\mathcal{F}$ to respect the $\mathbb{N}^d$-grading. For an $\alpha \in \mathbb{N}^d$ we denote by $S(-\alpha)$ the free S-module of rank 1 whose multigrading differs by α from the standard multigrading; that is for $\gamma \in \mathbb{N}^d$ the monomial $\mathbf{x}^\gamma \in S(-\alpha)$ has multidegree $\alpha + \gamma$. Also we require that the maps ∂_i are homogeneous with respect to this multigrading. Again it is well known (see again [18, Chapter

20]) that there is an $\mathbb{N}^d$-graded or multigraded minimal free resolution of $k[\Delta]$ – more generally of any $\mathbb{N}^d$-graded module over $k[x_1, \ldots, x_d]$. This leads to

$$\mathcal{F} : 0 \xrightarrow{\partial_{r+1}} \bigoplus_{\alpha \in \mathbb{N}^d} S(-\alpha)^{\beta_{r,\alpha}} \xrightarrow{\partial_r} \cdots \xrightarrow{\partial_1} \bigoplus_{\alpha \in \mathbb{N}^d} S(-\alpha)^{\beta_{0,\alpha}}$$

The numbers $\beta_{i,\alpha}$ determined by (multigraded) minimal free resolutions of $k[\Delta]$ are called the multigraded Betti numbers of $k[\Delta]$.

Example 1.2.10. Even though it is possible to describe the multigraded minimal free resolution of $k[\Delta]$ for $\Delta = U_{d,l}$ explicitly for all parameters j (see Exercise 1.4.13), we will confine ourselves in this example to the extreme cases $l = 0$ and $l = n - 1$.

- **l = 0**: For $U_{d,0}$ the Stanley-Reisner ring

$$k[U_{d,0}] = k[x_1, \ldots, x_d]/\langle x_1, \ldots, x_d \rangle \cong k$$

is the field k. We set $e_A = \sum_{i \in A} e_i$ and

$$F_i = \bigoplus_{\substack{A \subseteq [n] \\ \#A = i}} S(-e_A)$$

and define ∂_i for $f \in S(-e_A)$ and $A = \{j_0 < \cdots < j_{i-1}\}$ by

$$\partial_i f = \sum_{m=0}^{i-1} x_{j_m} f_{A \setminus \{j_m\}}.$$

Here we denote by $f_{A \setminus \{j_m\}}$ the element $f \in S(-e_A)$ considered as an element of $S(-e_{A \setminus \{j_m\}})$ with degree shifted by $-e_{j_m}$. Moreover, we deduce from the Auslander-Buchsbaum theorem that $\operatorname{depth}(k[U_{d,0}]) = 0 = \dim(U_{d,0}) + 1$. One checks that this indeed is a resolution and that since no units occur as coefficients in the differential it follows by (Min') that the resolution is minimal. This resolution actually will be one of the first examples of a cellular resolution that we will encounter in later chapters.

- **l = d − 1**: In this case

$$k[U_{d,d-1}] = k[x_1, \ldots, x_d]/\langle x_1 \cdots x_d \rangle.$$

We set $F_0 = S(-\underline{0})$ and $F_1 = S(-e_{[d]})$ where ∂_1 is defined by sending $1 \in S(-e_{[d]})$ to $x_1 \ldots x_d \in S(-\underline{0})$. Again one checks that this defines a resolution and then it is clear that the resolution is minimal. Again using the Auslander-Buchsbaum theorem we deduce that $\operatorname{depth}(k[U_{d,d-1}]) = d - 1 = \dim(U_{d,d-1}) + 1$.

Consider the generating series $H(S(-\alpha), t)$ of the monomials $\underline{\mathbf{x}}^\gamma$ in $S(-\alpha)$ by degree $|\alpha + \gamma|$. As before we write $|\cdot|$ for the coordinate sum of a vector in $\mathbb{N}^d$. Since the degrees in $S(-\alpha)$ differ only by a shift of $|\alpha|$ from the degrees in S we get $H(S(-\alpha), t) = \frac{t^{|\alpha|}}{(1-t)^d}$. Since $\mathcal{F}$ is exact it follows that

$$\sum_{i=0}^{r} (-1)^i \sum_{\alpha \in \mathbb{N}^d} \beta_{i,\alpha} H(S(-\alpha), t) = \mathrm{Hilb}_\Delta(t).$$

This implies that

$$\frac{\sum_{i=0}^{r} (-1)^i \sum_{\alpha \in \mathbb{N}^d} \beta_{i,\alpha} t^{|\alpha|}}{(1-t)^d} = \frac{\sum_{i=0}^{\dim(\Delta)+1} h_i t^i}{(1-t)^{\dim(\Delta)+1}}.$$

Hence we get an intimate relation between the multigraded Betti numbers of $k[\Delta]$ and the h-vector of $k[\Delta]$.

1.3 Cohen-Macaulay Simplicial Complexes

After having introduced some of the basic invariants of simplicial complexes that are studied in Algebraic Combinatorics, we now turn to a class of simplicial complexes that have particularly nice properties.

In Section 1.2 we had already used depth in order to describe the length of a minimal free resolution of $k[\Delta]$. Now we give a precise definition of depth and will use this invariant in order to define the Cohen-Macaulay simplicial complexes .

For a k-algebra R the depth depth(R) of R is the maximal length s of a sequence $y_1, \ldots, y_s$ of elements of R such that y_{i+1} is not a zero-divisor in $R/\langle y_1, \ldots, y_i \rangle$, $-1 \leq i \leq s-1$, and $0 \neq R/\langle y_1, \ldots, y_s \rangle$. Such a sequence is also called regular sequence. Indeed it can be shown that the elements of the sequence can be chosen to be homogeneous of degree 1 – after possibly extending the field k (see Exercise 1.4.19). Moreover, it can be shown that any permutation of the sequence is again a regular sequence (see Exercise 1.4.18).

Example 1.3.1. If $\Delta = U_{d,l}$ for $l = 0$ then $k[U_{d,0}] \cong k$ and depth$(k[U_{d,0}]) = 0$, since for any element y_1 of $k \setminus \{0\}$ we have $\langle y_1 \rangle = k$. If $\Delta = U_{d,l}$ for $l = d-1$ then $k[U_{d,d-1}] \cong k[x_1, \ldots, x_d]/\langle x_1 \cdots x_d \rangle$. In this case one checks that $y_1 = x_1 + x_2, \ldots, y_{d-1} = x_1 + x_d$ is a maximal regular and depth$(k[U_{d,d-1}]) = d-1$. Recall that in both cases we had already calculated the depth using the Auslander-Buchsbaum theorem in Example 1.2.10.

It is usually very hard to calculate the depth of a k-algebra and the situation does not become significantly simpler if we are in the combinatorial setting of $R = k[\Delta]$.

It turns out that in Combinatorics the situation when depth$(k[\Delta])$ = dim $\Delta + 1$ is particularly interesting. In this situation Δ is called Cohen-Macaulay over k. For those familiar with the notion of a Cohen-Macaulay ring in Commutative Algebra: The simplicial complex Δ is Cohen-Macaulay over the field k if and only if $k[\Delta]$ is a Cohen-Macaulay ring. This is due to the fact that in commutative algebra a k-algebra R is called Cohen-Macaulay if depth(R) = dim R (see [18]). Here dim R denotes the Krull-dimension of R. Note, depth$(R) \leq$ dim R is true always. The fact that the ring theoretic dimension dim $k[\Delta]$ equals dim $\Delta + 1$ then explains the combinatorial definition.

Example 1.3.2. If $\Delta = U_{d,l}$ for $l = 0$ then by Example 1.2.10 or 1.3.1 we know that

$$\text{depth}(k[U_{d,0}]) = 0 = -1 + 1 = \dim U_{n,0} + 1,$$

If $\Delta = U_{d,l}$ for $l = d - 1$ then again by Example 1.2.10 or 1.3.1 we know

$$\text{depth}(k[U_{d,d-1}]) = d - 1 = d - 2 + 1 = \dim U_{d,d-1} + 1.$$

Thus in both cases $k[\Delta]$ is Cohen-Macaulay over any field.

Note, that there are examples of simplicial complexes Δ for which the Cohen-Macaulay-ness of $k[\Delta]$ depends on k (see Exercise 1.4.17). From the side of Combinatorics the interest in Cohen-Macaulay simplicial complexes is stimulated by the following fact:

Proposition 1.3.3. *If Δ is a Cohen-Macaulay complex over some field k then its h-vector has only nonnegative entries.*

The proof of this result follows by some basic commutative algebra arguments from the definition of a Cohen-Macaulay ring (see Exercise 1.4.21).

It is easily seen that the converse of Proposition 1.3.3 is false (see Exercise 1.4.21). Given the definition of a Cohen-Macaulay simplicial complex it seems to be extremely hard to verify Cohen-Macaulay-ness for an even moderately complicated class of simplicial complexes.

Here we are at a point where Topological Combinatorics enters the picture. Let us first recall some basic notions from convex geometry. The convex hull conv$\{v_0, \ldots, v_m\}$ of a finite set of points $v_0, \ldots, v_m \in \mathbb{R}^j$ is the set of all convex linear combinations

$$\text{conv}\{v_0, \ldots, v_m\} := \Big\{ \sum_{i=0}^{m} \lambda_i v_i \ \Big| \ \lambda_i \geq 0, \ 0 \leq i \leq m, \ \sum_{i=0}^{m} \lambda_i = 1 \Big\}.$$

If $v_0, \ldots, v_m$ are affinely independent (i.e. $v_0 - v_m, \ldots, v_{m-1} - v_m$ are linearly independent) then their convex hull is called an (affine) m-simplex . Clearly, if V is any subset of an affinely independent set $v_0, \ldots, v_m$ then $\operatorname{conv}(V)$ is a $(\#V - 1)$-simplex – we call $\operatorname{conv}(V)$ a face of $\operatorname{conv}\{v_0, \ldots, v_m\}$. Faces that are 0-simplices are also called vertices and are identified with the unique point contained in them.

A geometric simplicial complex Γ in $\mathbb{R}^j$ is a collection of affine simplices in $\mathbb{R}^j$ such that for any simplex in Γ all its faces also lie in Γ and for any two simplices in Γ their intersection is a face of both. Clearly, if we map the geometric simplices in Γ to the set of their vertices then the image of Γ will be an abstract simplicial complex. Conversely, for a simplicial complex Δ on $[d]$ a geometric realization of Δ is a subset X – equipped with the subspace topology – of some $\mathbb{R}^j$ such that there is a geometric simplicial complex Γ for which the set theoretic union of its simplices is X. Since one can show that any two geometric realizations of an abstract simplicial complex Δ are homeomorphic we write $|\Delta|$ for any geometric realization of the abstract simplicial complex Δ on $[d]$. Indeed, if we take

$$|\Delta| = \bigcup_{A \in \Delta} \operatorname{conv}\{e_i \mid i \in A\},$$

where the e_i are the standard unit vectors e_i in $\mathbb{R}^d$, then it is a geometric realization of Δ. Let us state the upshot of this paragraph:

Proposition 1.3.4 (see e.g. [37]). *For any abstract simplicial complex Δ there is a geometric realization $|\Delta|$ and any two geometric realizations are homeomorphic.*

The preceding proposition allows us to give up the distinction between abstract and geometric simplicial complexes and their topological realizations for the rest of these notes.

Let us return to algebra and consider the dependence of $k[\Delta]$ on the topological realization $|\Delta|$. In general, from the point of view of topology the ring $k[\Delta]$ is 'bad' since it is not a topological invariant.

Example 1.3.5. For uniform matroids $U_{d,l}$ the geometric realization $|U_{d,l}|$ is just the $(l-1)$-skeleton of the $(d-1)$-simplex $\operatorname{conv}\{e_i \mid 1 \le i \le d\}$. In particular, for $l = 1$ we get d points and for $l = 2$ we get the geometric picture of a complete graph on d vertices. For $l = d - 1$ we get the full boundary complex of the $(d-1)$-simplex, which is easily seen to be homeomorphic to a $(d-2)$-sphere. If $d = 3$ then $|U_{3,2}|$ is just the geometric picture of a triangle. Clearly, as a topological space a triangle is homeomorphic to a quadrangle $|Q_4|$ – each homeomorphic to a 1-sphere. We can take for Q_4 the simplicial complex on $[4]$

with facets $\{1,2\}, \{2,3\}, \{3,4\}, \{1,4\}$. Then $k[U_{3,2}] = k[x_1, x_2, x_3]/\langle x_1 x_2 x_3 \rangle$ and $k[Q_4] = k[x_1, x_2, x_3, x_4]/\langle x_1 x_3, x_2 x_4 \rangle$. If we compute the Hilbert-series, we get

$$H_{U_{3,2}}(t) = \frac{(1 + t + t^2)}{(1 - t)^2}, H_{Q_4}(t) = \frac{(1 + 2t + t^2)}{(1 - t)^2},$$

being fundamentally different.

Thus $k[\Delta]$ clearly captures much of the combinatorial information of the particular triangulation of the space $|\Delta|$.

Nevertheless, it is surprising and worthwhile to mention that we can read off the Cohen-Macaulay property from the space $|\Delta|$.

Theorem 1.3.6 (Munkres [36]). *If Δ and Δ' are abstract simplicial complexes such that $|\Delta|$ and $|\Delta'|$ are homeomorphic then Δ is Cohen-Macaulay over k if and only if Δ' is Cohen-Macaulay over k.*

Since an algebraic approach to the Cohen-Macaulay property involves the calculation of the depth it is desirable to have a geometric criterion for Cohen-Macaulay-ness. Before we can state a theorem giving a criterion for Cohen-Macaulay-ness, we have to introduce some further notion from Topological Combinatorics.

For a simplicial complex Δ and a face $A \in \Delta$ we denote by $\text{link}_\Delta(A)$ the simplicial complex

$$\text{link}_\Delta(A) = \{B \subseteq [n] \setminus A \mid A \cup B \in \Delta\}.$$

Example 1.3.7. If $\Delta = U_{d,l}$ and $A \in \Delta$ then

$$\text{link}_\Delta(A) = \{B \subseteq [d] \setminus A \mid \#B \le l - \#A \}.$$

Thus up to relabeling the ground set elements we get that $\text{link}_\Delta(A)$ is an $U_{d-\#A, l-\#A}$.

If Δ is a simplicial complex on ground set $[d]$ then we denote by C_i the k-vectorspace of dimension f_i with basis f_A for $A \in \Delta$ and $\#A = i + 1$. By δ_i we denote the homomorphism sending f_A to $\sum_{j=0}^{i}(-1)^j f_{A \setminus \{v_j\}}$ if $A = \{v_0 < \cdots < v_i\}$. One easily checks that $\delta_i \circ \delta_{i+1} = 0$ and therefore $\text{Ker}\delta_i \subseteq \text{Im}\delta_{i+1}$. Then the i-th reduced simplicial homology group $\tilde{H}_i(\Delta; k)$ of Δ is defined as the quotient $\text{Ker}\delta_i/\text{Im}\delta_{i+1}$.

The criterion for Cohen-Macaulay-ness is provided by the following theorem of Reisner:

Theorem 1.3.8 (Reisner's Criterion [49]). *A simplicial complex Δ is Cohen-Macaulay over the field k if and only if for any $A \in \Delta$ we have $\tilde{H}_i(\text{link}_\Delta(A); k) = 0$ for $i \le \dim(\text{link}_\Delta(A)) - 1$.*

The following corollary shows that the concept of a pure simplicial complex – that we have introduced above but have not used so far – resurfaces here again.

Corollary 1.3.9. *If Δ is Cohen-Macaulay over some field k then Δ is a pure simplicial complex.*

Example 1.3.10.

1. If $\Delta = U_{d,l}$ then the image of δ_{i+1} for $i \leq l-2$ is equal to $\mathrm{Ker}\delta_i$. A basis for both is provided by the $\partial_{i+1}(f_B)$, where $n \in B$. Thus for $A = \emptyset$ we get

$$\widetilde{H}_i(\mathrm{link}_\Delta(\emptyset); k) = \widetilde{H}_i(U_{d,l}; k) = 0$$

   for $i < \dim U_{d,l} = l-1$. Since we already know that $\mathrm{link}_{U_{d,l}}(B) = U_{d-\#B, l-\#B}$ it follows by induction that $U_{d,l}$ is Cohen-Macaulay over any field k.

2. Let Δ be a simplicial complex such that $|\Delta|$ is homeomorphic to a D-sphere. Then one checks that $|\mathrm{link}_\Delta(A)|$ is homeomorphic to a $(D - \#A)$-sphere. Now standard arguments from Algebraic Topology [23] show that

$$\widetilde{H}_i(\mathrm{link}_A(\Delta); k) = 0, i < \dim(\mathrm{link}_A(\Delta)) = D - \#A.$$

   Thus Δ is indeed Cohen-Macaulay over any field.

Now we return to Enumeration. By Remark 1.2.8 (1) we know that $f_{\dim(\Delta)}$ is the sum over the h-vector. If Δ is Cohen-Macaulay then we know that $h_i \geq 0$ for all i. Thus the h_i give a statistics on the number of $\dim(\Delta)$-dimensional faces of Δ in that case. Recall that by Corollary 1.3.9 we know that all facets of a Cohen-Macaulay complex have the same dimension; that is Δ is pure.

Example 1.3.11. We have already seen that if $\Delta = U_{d,l}$ then Δ is Cohen-Macaulay and we know that in this case $\dim U_{d,l} = l-1$ and $f_{l-1} = \binom{d}{l}$. Thus the h_i provide statistics on the number of l-subsets of a d-set (see Exercise 1.4.16).

This approach finally raises the question which sequences of numbers can actually occur as the h-vector of a Cohen-Macaulay simplicial complex. In order to answer this question we again recall the binomial expansion from Lemma 1.2.5. So if for natural numbers n and i the binomial expansion of n is given by $n = \binom{a_i}{i} + \cdots \binom{a_j}{j}$ then we set

$$n^{\langle i \rangle} = \binom{a_i - 1}{i - 1} + \cdots + \binom{a_j - 1}{j - 1}.$$

Theorem 1.3.12 (see [12]). *A vector $(h_0, \ldots, h_D)$ of nonnegative numbers is the h-vector of a $(D-1)$-dimensional Cohen-Macaulay simplicial complex if and only if $h_0 = 1$ and $h_i^{\langle i \rangle} \leq h_{i-1}$ for $1 \leq i \leq D$.*

A standard proof of Theorem 1.3.12 is algebraic in nature and uses Exercise 1.4.19 and the famous Macaulay-Theorem about the Hilbert-series of 0-dimensional graded k-algebras, which says that a polynomial $h_0 + h_1 t + \cdots + h_D t^D$ is the Hilbert-series of a 0-dimensional k-algebra if and only if the h_i satisfy the conditions from Theorem 1.3.12.

Now having seen how enumerative invariants of Cohen-Macaulay graded k-algebras emerge and relate to enumerative invariants of simplicial complexes, the reader may ask how this relates to minimal free resolutions – one of the central concepts of these notes.

For that we need to define the concept of an Alexander-dual simplicial complex. For a simplicial complex Δ on ground set $[d]$ we define its (Alexander-) dual simplicial complex Δ^* as

$$\Delta^* := \{A \subseteq [d] \mid [d] \setminus A \notin \Delta\}.$$

We refer to Exercise 1.4.22 for more properties of Δ^*.

Now we are in position to formulate a result that gives a criterion for Cohen-Macaulay-ness of a simplicial complex in terms of the minimal free resolutions of $k[\Delta^*]$.

Theorem 1.3.13 (Reiner-Eagon Criterion [17]). *A simplicial complex Δ is Cohen-Macaulay over k if and only if the minimal $\mathbb{N}^d$-graded free resolution*

$$\mathcal{F} : 0 \xrightarrow{\partial_{r+1}} \bigoplus_{\alpha \in \mathbb{N}^d} S(-\alpha)^{\beta_{r,\alpha}} \xrightarrow{\partial_r} \cdots \xrightarrow{\partial_1} \bigoplus_{\alpha \in \mathbb{N}^d} S(-\alpha)^{\beta_{0,\alpha}}$$

of $k[\Delta^*]$ satisfies: There is an integer j such that for all $l \geq 1$ and $\alpha \in \mathbb{N}^d$ we have $\beta_{l,\alpha} = 0$ for $|\alpha| \neq l + j$.

Here we close the presentation of material linking Algebraic Combinatorics, Enumerative Combinatorics, Commutative Algebra and Algebraic Topology. As already mentioned we recommend the books [35], [57], [26] and the survey article [9] for further motivation and a broader view of the subject. In the last section of this chapter we present a small subset of the open problems in the field and add some speculations about their connections.

1.4 Some Open Problems in the Field

In this section we describe some of the important open problems in the field and add some questions and speculations. From now on we will be less self contained and will refer the reader to standard books or subsequent chapters for definitions and details.

The g-Conjecture:

First we discuss the *g*-conjecture. The *g*-conjecture asks for the classification of the *f*-vectors of simplicial complexes that triangulate a sphere or more generally simplicial complexes that are Gorenstein*. A simplicial complex Δ is called Gorenstein* over a field k if Δ is Cohen-Macaulay over k and for all simplices $B \in \Delta$ we have $\tilde{H}_i(\text{link}_\Delta(B); k) = k$ for $i = \dim(\text{link}_\Delta(B))$. Note that in particular this implies that Δ has the homology of a sphere and that it is a pseudomanifold (see Exercise 1.4.23). For this reason Gorenstein* simplicial complexes are sometimes also called generalized homology spheres. We would like to mention that the term homology sphere itself is used in various meanings in the literature. It can mean a space (manifold) that has the homology of a sphere or may be used synonymously to generalized homology sphere. The conjecture is formulated in terms of the *h*-vector (resp. the *g*-vector).

Conjecture 1.4.1 (g-Conjecture). A vector $(h_0, \ldots, h_D)$ of nonnegative integers is the *h*-vector of a $(D-1)$-dimensional Gorenstein* simplicial complex Δ if and only if

1. $h_i = h_{D-i}$ for $0 \leq i \leq D$.
2. $h_0 = 1$.
3. For $g_0 = 1$, $g_i = h_i - h_{i-1}$, $1 \leq i \leq \lfloor \frac{D}{2} \rfloor$ we have $g_{i-1} \geq g_i^{\langle i \rangle}$ for $1 \leq i \leq \lfloor \frac{D}{2} \rfloor$.

The conjecture is wide open but has been verified in the very important case for which it was originally conjectured. In the 1970's McMullen formulated the above conjecture for boundary complexes of simplicial polytopes. Billera and Lee [8] proved by explicit constructions the sufficiency of the conditions and Stanley [56] showed the necessity by invoking the Hard-Lefschetz Theorem from Algebraic Geometry. Later McMullen [34] gave a proof of the necessity without resorting to tools outside Discrete Geometry. We refer the reader for the basic definitions from polytope theory and further information about the *g*-Conjecture to the book by Ziegler [63]. In the sequel we will freely use polytope terminology without explicitly referring again to the book.

Theorem 1.4.2 (g-Theorem [8], [56]). *A vector $(h_0, \ldots, h_D)$ of nonnegative integers is the h-vector of the boundary complex of a D-dimensional simplicial polytope if and only if*

1. $h_i = h_{D-1}$ for $0 \leq i \leq D$.
2. $h_0 = 1$.
3. For $g_0 = 1$, $g_i = h_i - h_{i-1}$, $1 \leq i \leq \lfloor \frac{D}{2} \rfloor$ we have $g_{i-1} \geq g_i^{\langle i \rangle}$ for $1 \leq i \leq \lfloor \frac{D}{2} \rfloor$.

The g-Conjecture 1.4.1 remains wide open and opinions are split on whether or not there is a counterexample. Note that the sufficiency of the conditions in the g-Conjecture follows from the g-Theorem; it is the necessity that is open. Conditions (1) and (2) are easily verified, so that (3) becomes the crucial part. Recently some progress was made from a new perspective. Using rigidity arguments Nevo [39] was able to show that g_0, g_1, g_2 always satisfy (3), which implies the g-conjecture in dimensions ≤ 4. Since the conjectured characterization of f-vectors (resp. h-vectors) of boundary complexes of simplicial polytopes and triangulations of spheres is the same one may wonder how much larger the class of simplicial complexes that triangulate d-spheres is compared to those that appear as boundary complexes of d-dimensional simplicial polytopes. By Steinitz's Theorem (see [63, Theorem 4.1]) the two classes coincide in dimensions ≤ 2. But already in dimension 3 it is shown in [45] that the class of simplicial complexes that triangulate d-spheres is substantially richer.

h-Vectors of Gorenstein Algebras:

One way to derive conditions (1) and (2) of Conjecture 1.4.1 is via Commutative Algebra. For that we first have to note that if Δ is a Gorenstein* simplicial complex over the field k then $k[\Delta]$ is a standard graded Gorenstein k-algebra (see [12, Chapter 5]) . More generally, assume that $R = \bigoplus_{i \geq 0} R_i$ is a standard graded Gorenstein k-algebra (see [12, Chapter 3] or [57] for a definition and background). Then its Hilbert-series $\mathrm{Hilb}_R(t) := \sum_{i \geq 0} \dim R_i t^i$ satisfies

$$\mathrm{Hilb}_R(t) = \frac{h_0 + \cdots + h_D t^D}{(1-t)^{\dim R}},$$

where $\dim R$ is the Krull-dimension of R and $(h_0, \ldots, h_D)$ is a sequence of nonnegative integers – rationality follows by the Hilbert-Serre theorem (see [12, Proposition 4.3.3], [57]) and the nonnegativity of the h_i's from the fact that Gorenstein algebras are also Cohen-Macaulay (see [12, Theorem 3.2.10]) for which the result follows analogous to the case of Stanley-Reisner rings treated in Exercise 1.4.21. Note, that $\mathrm{Hilb}_{k[\Delta]}(t)$ coincides with $\mathrm{Hilb}_\Delta(t)$ from Section 1.2. The vector $(h_0, \ldots, h_D)$ is also called the h-vector of R. Again by Gorenstein-ness of R it follows that $(h_0, \ldots, h_D)$ satisfies $h_i = h_{D-i}$. But (3) remains open in general and is not true for Gorenstein k-algebras in general as shown in [40]. Indeed Stanley in [57, Section II.6] asks for a characterization of h-vectors of Gorenstein k-algebras und provides in [55, Example 4.1] an example which shows that (3) is not satisfied by h-vectors of Gorenstein algebras in general. Indeed he gives an example of a 0-dimensional Gorenstein k-algebra whose h-vector is not unimodal. A sequence of integers $(a_0, \ldots, a_r)$

is called unimodal if there is a $0 \le i \le r$ such that $a_0 \le \cdots \le a_i \ge \cdots \ge a_r$. It is a easy to see that if the g-conjecture holds then the h-vector of any Gorenstein* simplicial complex is unimodal. But there is still the following conjecture that is open.

Conjecture 1.4.3 (Hibi, Stanley). If $(h_0, \ldots, h_D)$ is the h-vector of a standard graded k-algebra that is a Gorenstein domain then it is unimodal.

There had been a conjecture by Hibi [25] on Cohen-Macaulay domains that would have implied Conjecture 1.4.3, but it was disproved [40]. Note that Conjecture 1.4.3 and the unimodality consequence of the g-Conjecture 1.4.1 are not directly related since the Stanley-Reisner ring of a Gorenstein* simplicial complex of dimension ≥ 0 is never a domain. In order to establish a link we turn to Gröbner bases theory. If $R = k[x_1, \ldots, x_d]/I$ is a standard graded k-algebra then I is a homogeneous ideal which means that it is generated by homogeneous polynomials. Gröbner bases theory now associates to I together with a monomial order a monomial ideal (see e.g. [22] for background on Gröbner basis theory). We briefly review the definitions needed in our context. A monomial order $\preceq$ on the set of monomials in the polynomial ring $k[x_1, \ldots, x_d]$ is a linear order for which $1 \preceq m$ for any monomial m and for which $m \preceq m'$ implies $mn \preceq m'n$ for any monomials m, m' and n. For a polynomial $f \in k[x_1, \ldots, x_d]$ we let $\mathrm{lt}_\preceq(f)$ be the largest monomial with respect to $\preceq$ that appears in f. For an ideal I the initial ideal with respect to $\preceq$ is the monomial ideal $\mathrm{in}_\preceq(I)$ generated by all monomials $\mathrm{lt}_\preceq(f)$ for $f \in I$. Now the following simple result (see Exercise 1.4.24) will be crucial for establishing a link between Conjecture 1.4.3 and the g-Conjecture 1.4.1.

Proposition 1.4.4. *For a standard graded k-algebra $k[x_1, \ldots, x_d]/I$ and a monomial order $\preceq$ we have*

$$\mathrm{Hilb}(k[x_1, \ldots, x_d]/I, t) = \mathrm{Hilb}(k[x_1, \ldots, x_d]/\mathrm{in}_\preceq(I), t).$$

Even though $\mathrm{in}_\preceq(I)$ is a monomial ideal it is in general not squarefree, the latter being equivalent to being the Stanley-Reisner ideal $\mathrm{in}_\preceq(I) = I_\Delta$ for some simplicial complex Δ (see Exercise 1.4.25). But still we get the following immediate corollary.

Corollary 1.4.5. *Let $R = k[x_1, \ldots, x_d]/I$ be a standard graded k-algebra and let $\preceq$ be a monomial order for which $\mathrm{in}_\preceq(I) = I_\Delta$ for some simplicial complex Δ. Then the h-vectors of R coincides with the h-vector of Δ (with terminal 0's removed).*

This of course is a rather trivial fact, but it allows us to connect the g-Conjecture 1.4.1 and Stanley's Conjecture 1.4.3.

Corollary 1.4.6. *Let $R = k[x_1, \ldots, x_d]/I$ be a standard graded Gorenstein domain and let Δ be a simplicial complex on ground set $[d]$. Assume there exists a monomial order $\preceq$ for which $\mathrm{in}_{\preceq}(I) = I_\Delta$. Then:*

1. *The h-vector of R is unimodal if and only if the h-vector of Δ is unimodal.*
2. *The h-vector of R satisfies conditions (1)-(3) of Conjecture 1.4.1 if and only if the h-vector of Δ satisfies conditions (1)-(3) of Conjecture 1.4.1.*

This again very simple observation has recently proved to be useful for many classes of Gorenstein domains. Indeed for a substantial number of classes of ideals I for which $k[x_1, \ldots, x_d]/I$ is a Gorenstein domain it has been verified that there is a monomial order $\preceq$ for which $\mathrm{in}_{\preceq}(I) = I_\Delta$ for a simplicial complex Δ whose geometric realization is a sphere. In this case we say that I has a spherical initial ideal . The first instance where it was shown that certain classes of ideals have a spherical initial ideal was joint work of Victor Reiner and the author in [48]. Here it was shown that the defining ideals of all Gorenstein Hibi-rings have a spherical initial ideal, indeed they show that there is an initial ideal that is the Stanley-Reisner ideal of a simplicial polytope. Gorenstein Hibi-rings include a toric deformation of coordinate rings of complex Grassmannians and quotients of the polynomial ring by some determinantal ideals. Later Athanasiadis [2] generalized the construction given in [48] to a much wider class of ideals. Both in [48], [2] and in further cases treated by Ohsugi and Hibi [43] the ideals were all toric. The work of Bruns und Römer [12] verified the existence of spherical initial ideals for all Gorenstein Ehrhart rings of integral polytopes which possess a regular unimodular triangulation. In all the cases mentioned so far it was possible to verify that indeed there is a simplicial polytope for which the Stanley-Reisner ideal occurs as an initial ideal. Together with the g-Theorem 1.4.2 this shows the following (we refer the reader to [13] for definition and further references):

Proposition 1.4.1 (Corollary 11 [13]) *If R is the Ehrhart ring of an integral polytope that admits a regular unimodular triangulation and R is Gorenstein then the h-vector of R satisfies (1)-(3) of Conjecture 1.4.1. In particular, it is unimodal.*

Beyond toric ideals in joint work with Jonsson [30] and Soll [54] we have exhibited (respectively conjectured) an explicit spherical initial ideal for the quotient of the polynomial ring in the entries of a generic skew-symmetric $(n \times n)$-matrix by its Pfaffians of fixed degree (respectively the quotient of the polynomial ring in the entries of a generic $(n \times n)$-matrix by its minors of a fixed size). Further results in this direction including results on minors of symmetric matrices and a very nice summary of the known examples can be found in the survey article [16]. We would like to mention that the latter article also exhibits interesting connections to minimal free resolutions.

All these results lead to the question:

Question 1.4.7. For which classes of standard graded k-algebras $k[x_1, \ldots, x_d]/I$ that are Gorenstein domains is there a spherical initial ideal of I ?

If the answer to Question 1.4.7 is positive and $\text{in}_{\prec}(I) = I_\Delta$ for a simplicial complex that is know to satisfy the g-Conjecture 1.4.1 it then it follows that R satisfies Stanley's Conjecture 1.4.3. Clearly, not all Gorenstein domains can be expected to even have a squarefree initial ideal. But even if there is no squarefree initial ideal one may wonder whether one of the operations that transform monomial ideals to squarefree monomial ideals will transform some initial ideal to a Stanley-Reisner ideal of a Gorenstein* simplicial complex.

It should be noted that the conditions of the g-Conjecture 1.4.1 also follow for a Gorenstein k-algebra if it can be shown that there is a quotient by a regular sequence which has the so called Lefschetz property (see e.g. [38]). We do not want to go into details here, but it should be noted that this path was taken by Stanley [56] in its original proof of the necessity in the g-Theorem 1.4.2. More recently, conditions on 0-dimensional algebras that imply the Lefschetz property have been studied intensively in Commutative Algebra (see e.g. [38]).

Betti Numbers of Resolutions:

So far our open problems centered around the h-vector. In this section we take a closer look at the Betti numbers of the minimal free resolution of a Stanley-Reisner ring. From now on we again write S for the polynomial ring $k[x_1, \ldots, x_d]$. At the end of Section 1.2 we have established the transformation between the multigraded Betti numbers β_{ij} of the S-module $k[\Delta]$ and the h-vector of Δ. Clearly in general, the Betti numbers cannot be calculated from the h-vector only. In particular, the classification theorems for f-vectors of simplicial complexes, Theorem 1.2.6, and h-vectors of Cohen-Macaulay simplicial complexes, Theorem 1.3.12, do not yield a classification of the sequence of Betti numbers of the corresponding class of simplicial complexes. Indeed such a classification is a wide open problem.

Question 1.4.8.   1. Is there a good characterization of the sequences of non-negative numbers $(\beta_0, \ldots, \beta_r)$ that can occur as the sequence of Betti numbers of a Stanley-Reisner ring $k[\Delta]$ of a simplicial complex Δ.

  2. Is there a good characterization of the matrices (β_{ij}) of nonnegative numbers that can occur as the bigraded Betti numbers of a Stanley-Reisner ring $k[\Delta]$ of a simplicial complex Δ.

Here the bigraded Betti number β_{ij} is defined as $\beta_{ij} := \sum_{\alpha \in \mathbb{N}^d, |\alpha|=j} \beta_{i,\alpha}$. For example, substantial work was needed [50] to only prove parts of a conjecture by Herzog on bounds on some of the bigraded Betti numbers. So in full generality this question, which indeed asks for a classification, appears to be almost hopeless.

For ideals I_Δ in S for which S/I_Δ is Gorenstein the minimal free resolution has additional structure which looks particularly intriguing in context of cellular free resolutions. This concept will become crucial in the later chapters and will be introduced in Section 2.4. For now it may suffice to say that a free resolution is cellular and supported by the CW-complex X if the differentials are the differentials from the cellular chain complex of X homogenized by monomial factors. Indeed, what we say in the next proposition about free resolutions of $k[\Delta]$ for Gorenstein* Δ is true for any Gorenstein S/I. But here we would like to confine ourselves to the case of Stanley-Reisner ideals and even more to Gorenstein* simplicial complexes. Also there are even more implications to the fact that S/I_Δ is Gorenstein (see Chapter 3 [12]).

Proposition 1.4.9 (Chapter 3 [12]). *Assume Δ is a Gorenstein* simplicial complex. Then there is a minimal free resolution*

$$\mathcal{F} : 0 \xrightarrow{\partial_{r+1}} F_r \xrightarrow{\partial_r} \cdots \xrightarrow{\partial_1} F_0$$

of I_Δ such that:

1. *$\beta_r = \text{rk}_S(F_r) = 1$.*
2. *$\beta_i = \text{rk}_S(F_i) = \text{rk}_S(F_{r-i-1}) = \beta_{r-i-1}$, $0 \leq i \leq r-1$.*
3. *$\partial_i = \partial_{r-i-1}^t$, $1 \leq i \leq r-1$, when considered as matrices.*

Now assume that for a Gorenstein* simplicial complex the minimal free resolution is cellular and supported by the CW-complex X. What do we know about X ? Clearly, Proposition 1.4.9 tells us that there is a unique cell of top dimension and that its removal from X leaves a CW-complex X' whose homology coincides with the one of a sphere of the same dimension. Assume X' is indeed a sphere and moreover that it is the boundary complex of a polytope. Then again Proposition 1.4.9 shows that the number of i-dimensional faces of X' equals the number of $(r-i-1)$-dimensional faces. Indeed, in addition $\partial_i = \partial_{r-i-1}^t$ also suggests that X' is indeed a self-dual polytope. A polytope is called self-dual if it has the same face structure as its polytope dual. Little is known about self-dual polytopes and there are not even many known classes (see for example [7]). Thus it appears to be a worthwhile undertaking to study cellular complexes that support the minimal free resolution of a Gorenstein* simplicial complex and find out whether these complexes can be chosen to be self-dual polytopes. This is rather daring for several reasons. First, it is known by work of Velasco [60] that not all monomial ideals even have a minimal free resolution that is supported by a CW-complex. Even though [60] does not explicitly clarify the Gorenstein* situation it at least makes it unlikely that in this case any minimal free resolution is cellular. Second, even if the minimal free resolution is supported by a CW-complex then by work of Reiner and the author [47] and more conceptually again by Velasco [60] we know that it may not be possible to choose a regular CW-complex. Recall, since a polytope is a regular CW-complex, regularity is a prerequisite for a CW-complex to be a polytope. Third, even if the monomial free resolution of the Stanley-Reisner ring of a Gorenstein* simplicial complex is supported by a regular CW-complex, then except for its homology there is little evidence that the CW-complex can be chosen to be a polytope. Last, even if it is supported by a polytope the duality of differential does not yet guarantee its self duality.

So let us be very modest and only ask for such constructions for very simple and nice Gorenstein* simplicial complexes; boundary complexes of simplicial polytopes.

Question 1.4.10. Let Δ be the boundary complex of a simplicial polytope. Is there a self-dual polytope that supports the minimal free resolution of I_Δ ?

Unfortunately, there are very few boundary complexes of simplicial polytopes for which we have a that good control over the minimal free resolution. Even the n-gon poses an open problem here.

Question 1.4.11. Is there a self-dual polytope that supports the minimal free resolution of the Stanley-Reisner ring of an n-gon ?

The last problem appears to be doable, even though work by Soll [53] shows that some obvious approaches do not work so easily. But again the work in [53] shows that the answer is yes for small n.

Here we close this introduction, hoping that we have given convincing evidence that the topic of the subsequent chapters relates well to central questions in Algebraic Combinatorics.

Exercises

Exercise 1.4.12. Give a proof of the binomial expansion Lemma 1.2.5.

Exercise 1.4.13. Describe for general parameters d and l the minimal free resolution of $k[U_{d,l}]$.

Exercise 1.4.14. Give a proof for $\widetilde{H}_{l-1}(U_{d,l}, k) \cong k^{\binom{d-1}{l}}$.

Exercise 1.4.15. Deduce Corollary 1.3.9 from Theorem 1.3.8.

Exercise 1.4.16. Find a statistics partitioning the set of l-subsets of $[d]$ into $l+1$ subsets $V_0, \ldots, V_l$. such that $(\#V_0, \ldots, \#V_l) = h(U_{d,l})$.

Exercise 1.4.17. Give an example of a simplicial complex Δ and fields k and k' such that Δ is Cohen-Macaulay over k but not over k'.

Exercise 1.4.18. Show that any permutation of a regular sequence is a regular sequence.

Exercise 1.4.19. Show that if $k[\Delta]$ is Cohen-Macaulay, then if k is sufficiently large there is a maximal regular sequence consisting of homogeneous polynomials of degree 1.

Exercise 1.4.20. Deduce Proposition 1.3.3 from Exercise 1.4.19.

Exercise 1.4.21. Give an example of a simplicial complex Δ that is not Cohen-Macaulay over any field (even any ring) but still has nonnegative h-vector.

Exercise 1.4.22. Let Δ be a simplicial complex over ground set Ω. Then:

1. Δ^* is a simplicial complex.
2. $(\Delta^*)^* = \Delta$.
3. The geometric realization of Δ^* is a deformation retract of the set theoretic difference of the boundary of the full simplex on Ω and the geometric realization of Δ. (This exercise requires some familiarity with basic homotopy theory).

Exercise 1.4.23. Show that any Gorenstein* simplicial complex is a pseudo-manifold (i.e. each face of codimension 1 is contained in exactly two facets).

Exercise 1.4.24. Prove Proposition 1.4.4. Hint: Show that the images of the monomials that do not lie in the initial ideal from a linear basis of both k-algebras.

Exercise 1.4.25. Show that a monomial ideal I is the Stanley-Reisner ideal of a simplicial complex if and only if it can be generated by squarefree monomials.

Basic Definitions and Examples

In this chapter we define the concept of a cellular resolution, which lays the ground for the application of discrete Morse theory in Commutative Algebra. Since the concept of a cellular resolution uses the concept of multigraded free resolutions and CW-complexes as an ingredient we define these concepts first. Indeed we have already given definitions of resolutions and simplicial complexes in Chapter 1 in a more informal way. In this chapter will be more rigorous and also add more details and examples. More precisely, Sections 2.1 - 2.4 give the definitions of free resolutions, CW-complexes, simplicial complexes, cellular free resolutions, monomial modules and co-Artinian monomial modules.

2.1 Multigraded Free Resolutions

We start with basic definitions from Commutative Algebra. In particular, we study multigraded free resolutions in more detail and give exact definitions. We first review some basic notions of Commutative Algebra and some of the definitions already provided in the introduction.

For a field k we denote by $k[x_1, \ldots, x_d]$ the polynomial ring over k in d variables. A monomial in $k[x_1, \ldots, x_d]$ is a product $\underline{x}^\alpha = x_1^{\alpha_1} \cdots x_d^{\alpha_d}$ for a vector $\alpha = (\alpha_1, \ldots, \alpha_d) \in \mathbb{N}^d$. The vector α is then called the multidegree of $\underline{x}^\alpha$ and the degree $\deg(\underline{x}^\alpha)$ is $|\alpha| = \alpha_1 + \cdots + \alpha_d$. Here we use $\mathbb{N}$ to denote the natural numbers including 0. As in Chapter 1 by $[d]$ we mean the set of integers $\{1, \ldots d\}$ form 1 up to d and for a subset $A \subseteq [d]$ we denote by $\underline{x}_A$ the monomial $\prod_{i \in A} x_i$.

Recall, that a finitely generated k-algebra R is up to isomorphism the quotient $k[x_1, \ldots, x_d]/I$ of the polynomial ring $k[x_1, \ldots, x_d]$ in a finite number of variables by an ideal I.

Example 2.1.1.

1. Let $k[x_1, \ldots . x_d]$ be the polynomial ring in d variables over the field k. Let $OS_{n,l}$ be the ideal that is spanned by all polynomials

$$\sum_{j=1}^{l} (-1)^j \mathbf{x}_{A \setminus \{i_l\}}$$

   for all subsets $A = \{i_1 < \cdots < i_l\} \subseteq [n]$ of cardinality $1 \le l \le d$. Then $k[x_1, \ldots, x_d]/OS_{d,l}$ is a finitely generated k-algebra. This ideal actually connects to matroids (see Example 1.2.1) and is known as the Orlik-Solomon ideal of the uniform matroid $U_{n,l-1}$ (see Question 2.1.14 for more details).
2. If we write $\mathfrak{m} = \langle x_1, \ldots, x_d \rangle$ for the ideal generated by the variables in $k[x_1, \ldots, x_d]$ then its l-th power $\mathfrak{m}^l$ is the ideal generated by all monomials $\mathbf{x}^\alpha$ of degree l; i.e. $|\alpha| = l$.

Definition 2.1.2.

1. *A finitely generated k-algebra R is called (standard) $\mathbb{Z}^d$-graded if there is a direct sum decomposition*

$$R = \bigoplus_{\alpha \in \mathbb{Z}^d} R_\alpha$$

   of R as a k-vectorspace by subspaces R_α, $\alpha \in \mathbb{Z}^d$, such that

$$R_\alpha R_\beta \subset R_{\alpha+\beta}$$

   and $R_{\mathbf{0}} = k$.
2. *If $R = \bigoplus_{\alpha \in \mathbb{Z}^d} R_\alpha$ is a $\mathbb{Z}^d$-graded k-algebra, then the R-module M is called $\mathbb{Z}^d$-graded if there is a direct sum decomposition*

$$M = \bigoplus_{\alpha \in \mathbb{Z}^d} M_\alpha$$

   as k-vectorspaces such that

$$R_\alpha M_\beta \subset M_{\alpha+\beta}.$$

   If

$$N = \bigoplus_{\alpha \in \mathbb{Z}^d} N_\alpha$$

   is another $\mathbb{Z}^d$-graded R-module and

$$\phi : M \longrightarrow N$$

   is an R-linear map, we call ϕ homogeneous if for all $\alpha \in \mathbb{Z}^d$ we have

$$\phi(M_\alpha) \subset N_\alpha.$$

Example 2.1.3.

1. Clearly the polynomial ring $R = k[x_1, \ldots, x_d]$ is $\mathbb{Z}^d$-graded. Here $R_\alpha = k\underline{x}^\alpha$ is the one dimensional vectorspace spanned by the monomial $\underline{x}^\alpha$ if $\alpha \in \mathbb{N}^d$ and $R_\alpha = 0$ otherwise. If I is any ideal in $k[x_1, \ldots, x_d]$ which is generated by monomials then the quotient $k[x_1, \ldots, x_d]/I$ inherits this $\mathbb{Z}^d$-grading and again is an example of a $\mathbb{Z}^d$-graded k-algebra.

2. As a special case of (1) the ideal $\mathfrak{m}^l$, which is generated by all monomials of degree l, inherits the $\mathbb{Z}^d$ grading from $k[x_1, \ldots, x_d]$ and therefore the quotient $k[x_1, \ldots, x_d]/\mathfrak{m}^l$ is $\mathbb{Z}^d$-graded as well.

3. If we denote by $T = k[x_1^{\pm 1}, \ldots, x_d^{\pm 1}]$ the ring of Laurent polynomials then T can also be considered as a $\mathbb{Z}^d$-graded $k[x_1, \ldots, x_d]$-module. The $\mathbb{Z}^d$-degree of a Laurent monomial $\underline{x}^\alpha = x_1^{\alpha_1} \cdots x_d^{\alpha_d}$ is given by $\alpha \in \mathbb{Z}^d$.

4. The k-algebra $k[x_1, \ldots, x_d]/OS_{n,l}$ in general carries no obvious $\mathbb{Z}^e$-grading for any $e \geq 2$. For any such grading we would need the generators of $OS_{n,l}$ to be homogeneous with respect to this grading on $k[x_1, \ldots, x_d]$. Still if we impose an $\mathbb{Z}$-grading on $k[x_1, \ldots, x_d]$ by the usual degree function, then all generators of $OS_{n,l}$ are homogeneous of degree $l - 1$. Thus the $\mathbb{Z}$-grading carries over to the quotient.

Let us now turn to a class of $\mathbb{Z}^d$-graded k-algebras that appears in the study of toric varieties (see e.g. [35] for more details on this connection).

Definition 2.1.4. *1. An affine semigroup Λ is a finitely generated (i.e., $\Lambda = \mathbb{N}\lambda_1 + \cdots + \mathbb{N}\lambda_r$ for some elements $\lambda_1, \ldots, \lambda_r \in \mathbb{N}^d$) sub-semigroup of the semigroup $\mathbb{N}^d$.*

2. For a given affine semigroup Λ we order $\mathbb{Z}^d$ by $\alpha \preceq \beta$ if and only if $\alpha + \gamma = \beta$ for some $\gamma \in \Lambda$. We write $(\mathbb{Z}^d, \Lambda)$ for the poset (i.e., partially ordered set) $\mathbb{Z}^d$ with the order induced by Λ.

3. Let k be a field. A ring R is called affine semigroup ring if $R = k[\Lambda]$ is the subring of $S := k[x_1, \ldots, x_d]$ with k-basis given by the monomials $\underline{x}^\lambda = x_1^{\lambda_1} \cdots x_d^{\lambda_d}$ for elements $\lambda = (\lambda_1, \ldots, \lambda_d) \in \Lambda$, where $\Lambda \subset \mathbb{N}^d$ is an affine semigroup.

4. Let $R = k[\Lambda]$ be an affine semigroup ring. We denote by $\mathfrak{m} \subset k[\Lambda]$ the maximal homogeneous ideal in $k[\Lambda]$. This is the ideal generated by all monomials $\underline{x}^\lambda = x_1^{\lambda_1} \cdots x_d^{\lambda_d}$ for elements $\lambda = (\lambda_1, \ldots, \lambda_d) \in \Lambda - \{(0, \ldots, 0)\}$.

Note, that an affine semigroup-ring $R = k[\Lambda]$ is naturally $\mathbb{Z}^d$-graded, where $R = \bigoplus_{\alpha \in \mathbb{Z}^d} R_\alpha$ with $R_\alpha = 0$ if $\alpha \notin \Lambda$ and $R_\alpha = k\underline{x}^\alpha$ if $\alpha \in \Lambda$. For $\Lambda = \mathbb{N}^d$ we have $k[\Lambda] = k[x_1, \ldots, x_d]$ and the partial order as defined above is the usual partial order given by componentwise comparison; that is $\alpha \preceq \beta$ if and only if $\alpha_i \leq \beta_i$ for all $i = 1, \ldots, d$. We will use the general setting of arbitrary Λ when we study resolutions of the field k over an affine semigroup ring $k[\Lambda]$.

Example 2.1.5. 1. If $\Lambda_l = \{\alpha \in \{0,1\}^d \mid |\alpha| = l\}$ for some $l \in \mathbb{N}\setminus\{0\}$ then Λ_l is called the *l-the squarefree Veronese* of $k[x_1,\ldots,x_d]$. If $l = 1$ then $\Lambda_1 = \mathbb{N}^d$ and $k[\Lambda_1] = k[x_1,\ldots,x_d]$. For $l = 2$ we get that $k[\Lambda_2] = k[x_i x_j \mid 1 \leq i < j \leq d]$. If we consider the polynomial ring $k[z_{i,j} \mid 1 \leq i < j \leq d]$ then the map sending $z_{i,j}$ to $x_i x_j$ defines a surjection onto $k[\Lambda_2]$. It can be seen that the kernel is the ideal generated by all binomials $z_{i,j}z_{r,s} - z_{i,s}z_{r,j}$. Thus if we grade $z_{i,j}$ be $e_i + e_j$ for the i-th and j-th unit vector in $\mathbb{N}^d$ then the polynomials $z_{i,j}z_{r,s} - z_{i,s}z_{r,j}$ are homogeneous of degree $e_i + e_j + e_r + e_s$ and the $\mathbb{Z}^d$ grading carries over to the quotient $k[\Lambda_2]$ and coincides with the $\mathbb{Z}^d$-grading inherited from $\Lambda_2 \subseteq \mathbb{N}^d$. If $d = 3$ and $l = 2$ then in the poset $(\mathbb{Z}^d, \Lambda_2)$ we get $(1,1,0) \prec (1,2,1)$ and $(0,1,1) \prec (1,2,1)$ but $(1,0,1) \not\prec (1,2,1)$ since $(0,2,0) \notin \Lambda_2$.

2. For a permutation $\sigma \in S_d$ we denote by $X^\sigma = (x_{ij}^\sigma) \in \mathbb{R}^{d \times d}$ the corresponding permutation matrix ; i.e. $x_{ij}^\sigma = 0$ for $j \neq \sigma(i)$ and $x_{i\sigma(i)}^\sigma = 1$. Then the set of all permutations matrices X^σ for $\sigma \in S_d$ generates an affine semigroup, which we denote by B_d. The corresponding affine semigroup ring is of particular combinatorial interest. The Hilbert-series $H(k[B_d], t) = \sum_{i \geq 0} a_i^d t^i$ has the property that the coefficients a_i^d count the number of magic squares – $d \times d$ matrices with entries in $\mathbb{N}$ – with row and column sums i. The numbers a_i^d have been subject to considerable attention in Enumerative Combinatorics and provided one of initial motivations for the introduction of techniques from Commutative Algebra into Combinatorics (see [57] for more background).

Definition 2.1.6 (Multigraded free chain complex). *Let R be a $\mathbb{Z}^d$-graded k-algebra, $\mathcal{C}_i$ for all $i \in \mathbb{N}$ a free (finitely generated) $\mathbb{Z}^d$-graded R-module and*

$$\mathcal{C}: \ldots \xrightarrow{\partial_{i+1}} \mathcal{C}_i \xrightarrow{\partial_i} \mathcal{C}_{i-1} \xrightarrow{\partial_{i-1}} \ldots \xrightarrow{\partial_1} \mathcal{C}_0$$

a sequence such that all maps $\partial_i : \mathcal{C}_i \longrightarrow \mathcal{C}_{i-1}$, $i \in \mathbb{N} - \{0\}$, are $\mathbb{Z}^d$-homogeneous and $\partial_i \circ \partial_{i+1} = 0$ for all $i \in \mathbb{N}-\{0\}$. Then $\mathcal{C}$ is called a $\mathbb{Z}^d$-graded or multigraded free chain complex (over R).

Example 2.1.7. Let $R = k[x_1,\ldots,x_d]$ and consider $\mathcal{C}: 0 \xrightarrow{\partial_2} \mathcal{C}_1 \xrightarrow{\partial_1} \mathcal{C}_0$, where

$$\mathcal{C}_1 = R(-(1,1,0)) \oplus R(-(1,0,1)) \oplus R(-(0,1,1)),$$

$$\mathcal{C}_0 = R(-(1,0,0)) \oplus R(-(0,1,0)) \oplus R(-(0,0,1)).$$

We define ∂_1 by sending $f \in R(-e_i - e_j)$ to $x_i f_i - x_j f_j$ for $1 \leq i < j \leq 3$, where f_i (resp. f_j) denotes $f \in R(-e_i - e_j)$ considered as an element of $R(-e_i)$ (resp. $R(-e_j)$). As usual e_i, e_j denote the corresponding unit vectors in $\mathbb{N}^3$. Then $\mathcal{C}$ is an $\mathbb{Z}^3$-graded chain complex.

Remark 2.1.8. Note that if $R = k[\Lambda]$ is an affine semigroup ring and

$$\mathcal{C}: \quad \ldots \xrightarrow{\partial_{i+1}} \mathcal{C}_i \xrightarrow{\partial_i} \mathcal{C}_{i-1} \xrightarrow{\partial_{i-1}} \ldots \xrightarrow{\partial_1} \mathcal{C}_0$$

is an $\mathbb{Z}^d$-graded free chain complex over R then $\mathcal{C}_i = \bigoplus_{\alpha \in \mathbb{Z}^d} C_i^\alpha$ where $C_i^\alpha = R(-\alpha)^{\beta_i^\alpha}$ is the free R-module of rank β_i^α with generators in multidegree α.

Definition 2.1.9 (Multigraded free resolution). *Let $R = k[\Lambda]$ be an affine semigroup ring. Let $M = \bigoplus_{\alpha \in \mathbb{Z}^d} M_\alpha$ be a $\mathbb{Z}^d$-graded R-module. A $\mathbb{Z}^d$-graded free resolution or multigraded free resolution $\mathcal{F}$ of M is a $\mathbb{Z}^d$-graded free chain complex*

$$\mathcal{F}: \quad \cdots \xrightarrow{\partial_{i+1}} \mathcal{F}_i \xrightarrow{\partial_i} \mathcal{F}_{i-1} \xrightarrow{\partial_{i-1}} \ldots \xrightarrow{\partial_1} \mathcal{F}_0$$

over R, satisfying the following properties:

1. $\mathcal{F}$ is exact, that is,

$$\mathrm{Ker}\partial_i = \mathrm{Im}\partial_{i+1} \text{ for all } i \in \mathbb{N} \setminus \{0\},$$

2. Coker $\partial_1 \cong M$.

For first examples of multigraded free resolutions we refer to the examples given in the introduction 1.2.10.

Remark 2.1.10. Let $R = k[\Lambda]$ be an affine semigroup ring and

$$\mathcal{F}: \quad \cdots \xrightarrow{\partial_{i+1}} \mathcal{F}_i \xrightarrow{\partial_i} \mathcal{F}_{i-1} \xrightarrow{\partial_{i-1}} \ldots \xrightarrow{\partial_1} \mathcal{F}_0$$

a $\mathbb{Z}^d$-graded free chain complex. Then $\mathcal{F}$ is a free resolution of the $\mathbb{Z}^d$-graded R-module M if and only if there is a homomorphism of $\partial_0 : \mathcal{F}_0 \to M$ of $\mathbb{Z}^d$-graded modules, such that

$$\cdots \xrightarrow{\partial_{i+1}} \mathcal{F}_i \xrightarrow{\partial_i} \mathcal{F}_{i-1} \xrightarrow{\partial_{i-1}} \ldots \xrightarrow{\partial_1} \mathcal{F}_0 \xrightarrow{\partial_0} M \to 0$$

is exact.

Definition 2.1.11 (Minimality of a multigraded free resolution). *A $\mathbb{Z}^d$-graded free resolution*

$$\mathcal{F}: \quad \cdots \xrightarrow{\partial_{i+1}} \mathcal{F}_i \xrightarrow{\partial_i} \mathcal{F}_{i-1} \xrightarrow{\partial_{i-1}} \ldots \xrightarrow{\partial_1} \mathcal{F}_0,$$

$\mathcal{F}_i = \bigoplus_{\alpha \in \mathbb{Z}^d} F_i^\alpha$, $F_i^\alpha = R(-\alpha)^{\beta_i^\alpha}$ of the $\mathbb{Z}^d$-graded $R = k[\Lambda]$-module M is called minimal if for all $i \in \mathbb{N} - \{0\}$ the image of ∂_i is contained in $\mathfrak{m}\mathcal{F}_{i-1}$. The numbers

$$\beta_i^\alpha(M) := \beta_i^\alpha$$

are called the (multigraded) Betti numbers of M. By $\beta_i = \sum_{\alpha \in \mathbb{Z}^d} \beta_i^\alpha$ we denote the Betti numbers of M.

Remark 2.1.12. Existence and uniqueness up to chain-complex isomorphisms of the minimal free resolution are well known facts (see [18]) from Commutative Algebra. The term *minimal* is derived from the fact that a minimal resolution

$$\cdots \xrightarrow{\partial_{i+1}} \mathcal{F}_i \xrightarrow{\partial_i} \mathcal{F}_{i-1} \xrightarrow{\partial_{i-1}} \cdots \xrightarrow{\partial_1} \mathcal{F}_0$$

of the R-module M simultaneously assumes minimal values for all ranks β_i^α, where $\mathcal{F}_i = \bigoplus_{\alpha \in \mathbb{Z}^d} F_i^\alpha$ and $F_i^\alpha = R(-\alpha)^{\beta_i^\alpha}$.

Exercises and Questions

Exercise 2.1.13. Given a minimal free resolution of the ideal I in $k[x_1, \ldots, x_d]$. Show how to construct from the given minimal free resolution a minimal free resolution of the quotient $k[x_1, \ldots, x_d]/I$. Note, the construction should transform multigraded resolutions into multigraded resolutions.

Question 2.1.14. Let M be a matroid over the ground set $[d]$ (see Example 1.2.1 for a definition of matroids). Recall, that a set $A = \{j_1 < \cdots < j_l\} \subseteq [d]$ is called a circuit of M if $A \notin M$ and $A \setminus \{j_i\} \in M$ for all $1 \le i \le l$ (see also Example 1.2.7). Let OS_M be the ideal in $k[x_1, \ldots, x_d]$ that is generated by all $\sum_{i=1}^l (-1)^i \mathbf{x}_{A \setminus \{j_i\}}$ for circuits $A = \{j_1 < \cdots < j_l\}$ of M (see Example 2.1.1 (1) for the case $M = U_{d,l}$). Then $k[x_1, \ldots, x_d]/OS_M$ is a commutative version of the Orlik-Solomon algebra (see [44]). The Orlik-Solomon ideal for general matroids has been subject to substantial research recently (see for example [58]) and still appears to be a mysterious and difficult object. In particular, the minimal free resolution of OS_M is not known for general M.

Question 2.1.15. Let M be a matroid (see Example 1.2.1) over $[d]$. Let us denote by $\{B_1, \ldots, B_f\}$ the collection of its inclusionwise maximal independent sets – the bases of the matroid M. Then consider the affine semigroup $\Lambda_M \subseteq \mathbb{N}^d$ that is generated by the vectors $e_{B_i} = \sum_{j \in B_i} e_j$ for the j-th unit vectors e_j in $\mathbb{R}^d$. It has been asked by Sturmfels whether the minimal free resolution of the field k – considered as an $\mathbb{Z}^d$-graded $k[\Lambda_M]$-module by $k \cong k[\Lambda_M]/\mathfrak{m}$ for the ideal $\mathfrak{m}$ generated by the x_{B_j}. – can be described. In particular, it is open whether $\beta_{i,\alpha} = 0$ for $|\alpha| \ne i$. The latter question is equivalent to saying that $k[\Lambda_M]$ is Koszul.

Exercise 2.1.16.

1. For $l \ge 1$ find a $\mathbb{Z}^d$-graded minimal free resolution of the ideal $\langle x_1^l, \ldots, x_d^l \rangle$ in $k[x_1, \ldots, x_d]$.
2. Find a $\mathbb{Z}^d$-graded minimal free resolution of the ideal $\langle x_{2i+1} x_{2i+2} \mid 0 \le i \le d-1 \rangle$ in $k[x_1, \ldots, x_{2d}]$.

2.2 Basics of CW-Complexes

After being prepared to work with free resolutions we will now recall some basic definitions from Algebraic Topology. In particular, we will give the definition and examples of CW-complexes. In later chapters this seemingly unrelated concept will allow us to endow free resolutions with a geometric structure.

Let us start with the basic definitions around CW-complexes. We refer the reader to the book [23] for more details.

Definition 2.2.1. *A topological space is called an (open) cell of dimension d (or d-cell) if it is homeomorphic to the d-dimensional open ball*

$$\overset{\circ}{B}{}^d = \{x = (x_1, \ldots, x_d) \in \mathbb{R}^d \mid \sum_{i=1}^{d} x_i^2 < 1\}.$$

Analogously, a topological space is called an closed cell of dimension d (or closed d-cell) if it is homeomorphic to the d-dimensional ball

$$B^d = \{x = (x_1, \ldots, x_d) \in \mathbb{R}^d \mid \sum_{i=1}^{d} x_i^2 \leq 1\}.$$

Definition 2.2.2 (CW-complex). *We call a topological space X a CW-complex, if there exists a collection $X^{(*)} = \{\sigma_i \mid i \in I\}$ of disjoint open cells such that*

$$X = \bigcup_{i \in I} \sigma_i,$$

satisfying the following properties:

1. *X is Hausdorff.*
2. *For every open cell $\sigma \in X^{(*)}$ of dimension d, there exists a continuous map*

$$f_\sigma : \sigma \subseteq B^d \longrightarrow X$$

from the corresponding closed d-cell B^d such that the restriction $f_\sigma^\circ := f_\sigma|_{\overset{\circ}{B}{}^d}$ is a homeomorphism

$$f_\sigma^\circ : \overset{\circ}{B}{}^d \overset{\cong}{\longrightarrow} \sigma$$

and such that the image $f_\sigma(S^{d-1})$ of the boundary S^{d-1} of B^d intersects only finitely many cells non-trivially, all of which have dimension at most $d-1$.
3. *A subset $A \subset X$ is closed in X if and only if $A \cap \bar\sigma$ is closed in $\bar\sigma$ for all $\sigma \in X^{(*)}$.*

For a cell $\sigma \in X^{(*)}$, we call the map

$$f_\sigma : B^d \longrightarrow X$$

the characteristic map of σ and $\bar{\sigma} = f_\sigma(B^d)$ the closed cell that belongs to σ.

For $d \in \mathbb{N}$ we denote by $X^d \subset X$ the union of all cells of dimension at most d and call X^d the d-skeleton of X. By $X^{(d)}$ we denote the set of all cells of dimension d.

A subcomplex of X is given by a subset of the cells of X that forms a CW-complex with the same characteristic maps. We say (X, A) is a pair of CW-complexes (or CW-pair) if X is a CW-complex and A a subcomplex of X.

If (X, A) is a pair of CW-complexes then we denote by $(X, A)^{(d)}$ the set of d-cells of X that do not belong to A. We set $(X, A)^{(*)} := \bigcup_{d \geq 0}(X, A)^{(d)}$.

There is a natural way to view $(X, A)^{(*)}$ ($X^{(*)}$ respectively) as a partially ordered set: For cells $\sigma', \sigma \in (X, A)^{(*)}$ ($X^{(*)}$ respectively) we set $\sigma' \leq \sigma$ if and only if the closed cell $\overline{\sigma'}$ is a subset of the closed cell $\bar{\sigma}$. We say that σ' is a facet of σ if $\sigma' \neq \sigma$ and for $\tau \in (X, A)^{(*)}$ the inclusion $\sigma' \leq \tau \leq \sigma$ implies $\tau \in \{\sigma, \sigma'\}$. We say that σ is a facet of X if σ is maximal with respect to the above partial relation on $X^{(*)}$.

Fig. 2.1 gives an example of a CW-structure on the 2-sphere.

Example 2.2.3. Since a (closed) 0-cell consists of a point only there is a unique map from any closed d-cell to a 0-cell. The resulting attaching map corresponds to collapsing the boundary of B^d to a point. Thus any CW-complex consisting of a 0-cell and a single d-cell for some $d \geq 1$ only is homeomorphic to a d-sphere S^d.

Remark 2.2.4. 1. Note that in all constructions mentioned above, a single CW-complex X can also be regarded as the CW-pair $(X, \emptyset)$. In the remainder of these lecture notes we will sometimes implicitly identify X with $(X, \emptyset)$.
2. The letters "CW" in the term "CW-complex" stand for "closure - finiteness" (referring to Property 2 in Definition 2.2.2.) and "weak topology" (referring to Property 3 in Definition 2.2.2.).

The Definition 2.2.2 of a CW-complex shows that indeed the data needed to define a CW-complex is sufficient to reconstruct this complex from the subcomplexes formed by its faces of dimension $\leq i$ for all $i \geq 0$. This constructive viewpoint is actually better suited for combinatorial application. The remaining section is devoted to a presentation of this aspect. First, we need to recall the process of gluing two topological spaces.

Skeleta   **Cells**

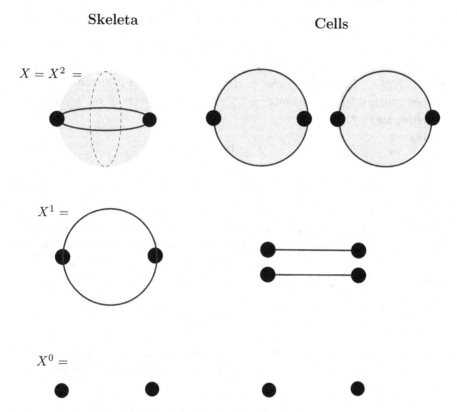

$X = X^2 =$

$X^1 =$

$X^0 =$

Fig. 2.1. Realizing the 2-sphere as a CW-complex

Definition 2.2.5. *If X and Y are topological spaces, $A \subset X$ closed, $f : A \longrightarrow Y$ a continuous map, we denote by $X \cup_f Y$ the quotient space of the disjoint union $X \dot\cup Y$ by the equivalence relation that is generated by the relations $a \sim f(a)$ for all $a \in A$. The space $X \cup_f Y$ is endowed with the quotient topology induced by this equivalence relation.*

Lemma 2.2.6. *A topological space X is a CW-complex if and only if it can be obtained by the following process of attaching cells of increasing dimensions:*

Let X^0 be a discrete set of points. These points form the 0-skeleton. Assume the $(d-1)$-skeleton X^{d-1} is constructed. Let $\dot\bigcup_{\sigma \in I_d} B_\sigma^d$ be a disjoint union of balls $B_\sigma^d \cong B^d$ of dimension d, σ ranging over some index set I_d. For all $\sigma \in I_d$ let $f_{\partial\sigma} : S^{d-1} = \partial B^d \longrightarrow X^{d-1}$ be a continuous map whose image intersect only finitely many cells in X^{d-1} non-trivially.. These maps induce a continuous map

$$f : \dot\bigcup_{\sigma \in I_d} S_\sigma^{d-1} = \partial(\dot\bigcup_{\sigma \in I_d} B_\sigma^d) \longrightarrow X^{d-1},$$

S_σ^{d-1} denoting the sphere that bounds B_σ^d. We construct the d-skeleton of X by setting

$$X^d := (\dot{\bigcup}_{\sigma \in I_d} B_\sigma^d) \cup_f X^{d-1}.$$

Note that the map f carries the information of how the d-dimensional balls B_σ^d are attached by their boundaries S_σ^{d-1} to X^{d-1}. The map $f_{\partial\sigma}$ is called the attaching map of σ.

We set

$$X := \bigcup_{d \in \mathbb{N}} X^d.$$

We define the topology of X by setting for $A \subset X$: A is closed in X if and only if $A \cap X^d$ is closed in X^d for all $d \in \mathbb{N}$.

Proof. If a topological space X is given by a construction as described above, the cells of X are in one-to-one correspondence with the elements of the disjoint union $I := \bigcup_{d \in \mathbb{N}} I_d$: The characteristic map f_σ of the cell corresponding to the index $\sigma \in I_d$ is given by the composition

$$B^d \xrightarrow{\cong} B_\sigma^d \hookrightarrow \dot{\bigcup}_{\sigma \in I_d} B_\sigma^d \longrightarrow (\dot{\bigcup}_{\sigma \in I_d} B_\sigma^d) \cup_f X^{d-1} \hookrightarrow X.$$

The cell itself is given by $f_\sigma(\mathring{B}^d)$. On the other hand, if X is a CW-complex and $\sigma \in X^{(*)}$ a cell, $f_\sigma : B^d \longrightarrow X$ its characteristic map, we define the corresponding attaching map by

$$f_{\partial\sigma} := f_\sigma|_{S^{d-1}} : S^{d-1} = \partial B^d \longrightarrow X^{d-1} :$$

Using the properties of the quotient map

$$\dot{\bigcup}_{\sigma \in I_d} B_\sigma^d \dot{\cup} X^{d-1} \longrightarrow (\dot{\bigcup}_{\sigma \in I_d} B_\sigma^d) \cup_f X^{d-1} ,$$

it is straightforward to check that the identity map between two copies of X, one endowed with the topology according to the definition of the CW-complex and one with that according to the attaching construction, is continuous in both directions. This proves the assertion.

Exercises

Exercise 2.2.7. Let $\mathcal{A} = \{H_1, \ldots, H_f\}$ be a finite set of affine hyperplanes in $\mathbb{R}^d$. Fix for each hyperplane H_j a linear form $\ell_j : \mathbb{R}^d \to \mathbb{R}$ such that $H_j = \mathrm{Ker}\,\ell_j$. To each point $\underline{x} \in \mathbb{R}^d$ we associate the sign-vector $\mathrm{sgn}(\underline{x}) = (\mathrm{sgn}(\ell_1(\underline{x})), \ldots, \mathrm{sgn}(\ell_f(\underline{x})))$, where for a real number $a \in \mathbb{R}$ its sign $\mathrm{sgn}(a) \in \{0, \pm 1\}$ is defined in the usual way. For each $\epsilon = (\epsilon_1, \ldots, \epsilon_f) \in \{0, \pm 1\}^f$ consider $c_\epsilon = \{\underline{x} \in \mathbb{R}^d \mid \mathrm{sgn}(\underline{x}) = \epsilon\}$. the collection of points in $\mathbb{R}^d$ with sign-vector ϵ.

1. Show that each c_ϵ is an open cell.
2. Let $X_\mathcal{A}$ be the union of the cells c_ϵ that are bounded subsets of $\mathbb{R}^d$. Show that $X_\mathcal{A}$ is a CW-complex with cell structure given by the bounded c_ϵ.

Exercise 2.2.8. Let S^1 be the unit circle. Construct for $n \geq 1$ a CW-complex X_n which is homeomorphic to $(S^1)^n$ and minimizes the number of cells among all CW-complexes homeomorphic to $(S^1)^n$.

2.3 Basics of Cellular Homology

In this section we introduce cellular homology for CW-complexes X and for pairs (X, A) of CW-complexes (see [11] or [23]). Indeed, it seems that the definition of cellular homology is much more complicated than singular homology – and of course it is well know that the two theories coincide on CW-complexes. But in Topological and Algebraic Combinatorics one cares about explicit constructions and calculations. Thus it is necessary to set up computable homology theories and cellular homology is one of them. Here we would like to be more precise about what we mean by a 'computable homology theory.' We will require that as input data the space is given in a form from which one can build chain groups with explicit bases and matrices representing the differentials. Cellular homology is computed from the cellular chain complex, which we now define.

Definition 2.3.1 (Cellular chain complex). *Let (X, A) be a CW-pair. For $i \in \mathbb{N}$ let $\mathcal{C}_i(X, A)$ be the k-vectorspace freely generated by the i-cells of X that do not lie in A. The following construction defines the differentials of a chain complex*

$$\mathcal{C}(X, A) : \quad \cdots \xrightarrow{\partial_{i+1}} \mathcal{C}_i(X, A) \xrightarrow{\partial_i} \mathcal{C}_{i-1}(X, A) \xrightarrow{\partial_{i-1}} \cdots \xrightarrow{\partial_1} \mathcal{C}_0(X, A) \xrightarrow{0} 0 :$$

called the cellular chain complex of the pair (X, A).

For $i \in \mathbb{N}$ let $\pi_i : X^i \to X^i/X^{i-1}$ be the canonical projection. Note that

$$X^i/X^{i-1} \cong \bigvee_{\sigma \in X^{(i)}} S_\sigma^i,$$

with $S_\sigma^i = S^i \times \{\sigma\}$ denoting a sphere representing the image of the closed cell $f_\sigma(B^i)$ under the projection π_i. Here the wedge is taken by choosing a base point $* \in S^i$ and by identifying all $(*, \sigma)$, $\sigma \in X^{(i)}$.

For $i \in \mathbb{N}$ let $\gamma_i : B^i \longrightarrow S^i$ be any continuous map collapsing the bordering sphere S^{i-1} of B^i to the base point of S^i such that the restriction to the interior of B^i is a homeomorphism onto its image. For $\sigma \in (X, A)^{(i)}$ define

$p_\sigma : X^i \to S^i$ such that $p_\sigma \circ f_\sigma = \gamma_i$ and $p_\sigma \circ f_{\sigma'}$ is the constant map onto the base point of S^i for all $\sigma' \in (X, A)^{(i)}$, $\sigma \neq \sigma'$. We call p_σ the cellular projection map corresponding to the cell σ.

Recall, that for a map $f : S^i \longrightarrow S^i$ one defines its degree $\deg f$ to be the unique integer such that the induced homomorphism in homology $f_* : \tilde{H}_i(S^i; \mathbb{Z}) \longrightarrow \tilde{H}_i(S^i; \mathbb{Z})$ is multiplication with $\deg f$.

Let σ be an i-cell. We set

$$\partial_i(\sigma) := \sum_{\tau \in X^{(i-1)}} [\tau : \sigma]\tau.$$

with

$$[\tau : \sigma] := \deg(p_\tau \circ f_{\partial\sigma}).$$

This defines the differential $\partial_i : \mathcal{C}_i(X, A) \longrightarrow \mathcal{C}_{i-1}(X, A)$.

Indeed Definition 2.3.1 bears the claim that the ∂_i are differentials; i.e. $\partial_i \circ \partial_{i+1} = 0$ for all $i \geq 0$. We leave the verification of this claim as Exercise 2.3.7. The next result now shows that the complex we have set up indeed calculates the 'right' homology theory.

Theorem 2.3.2 (see p. 204 [11]). *The singular homology $H_*(X, A)$ of the pair (X, A) is isomorphic to the homology of the chain complex $\mathcal{C}(X, A)$.*

For our applications to free resolutions the class of acyclic complexes will be central.

Definition 2.3.3 (Acyclic CW-complex). *Let k be a field. A CW-complex X is called acyclic if we have:*

$$H_i(X; k) = \begin{cases} k & \text{for } i = 0 \\ 0 & \text{otherwise.} \end{cases}$$

Note that a connected CW-complex X is acyclic if and only if $\text{Ker}\partial_i = \text{Im}\partial_{i+1}$ for $i \geq 1$. This condition is reminiscent of the same condition for the differential in a resolution. We we will make this more precise in Proposition 3.1.2.

CW-complexes form a generalization of spaces called *(geometric) simplicial complexes* . These are built from affine simplices in the following way:

Definition 2.3.4 (Simplicial Complex).

1. *Let $v_0, \ldots, v_i \in \mathbb{R}^d$ be a finite set of points in Euclidian space. The points $v_0, \ldots, v_i$ are called affinely independent if $\sum_{\nu=0}^{i} \lambda_\nu v_\nu = 0$ and $\sum_{\nu=0}^{i} \lambda_\nu = 0$ implies all $\lambda_\nu = 0$. (This is equivalent to the property that the smallest affine subspace of $\mathbb{R}^d$ which contains the points $v_0, \ldots, v_i$ has dimension i.)*

2. A (geometric) simplex $\sigma = (v_0, \ldots, v_i)$ is defined as the set of all linear combinations $\sum_{\nu=0}^{i} \lambda_\nu v_\nu$ such that $\sum_{\nu=0}^{i} \lambda_\nu = 1$ and all $\lambda_\nu \geq 0$, where $v_0, \ldots, v_i$ is a set of affinely independent points in some $\mathbb{R}^d$.

3. For a geometric simplex $\sigma = (v_0, \ldots, v_i)$ we refer to i as the dimension of σ.

4. We regard $() = \emptyset$ as the unique simplex of dimension -1.

5. Note that if $\sigma = (v_0, \ldots, v_i)$ is an geometric simplex then $(v_{\nu_0}, \ldots, v_{\nu_k})$ is a geometric simplex for any $0 \leq \nu_0 < \ldots < \nu_k \leq i$ as well. We call $(v_{\nu_0}, \ldots, v_{\nu_k})$ a k-face of σ.

6. A set Σ of geometric simplices is called simplicial complex if
   a) $\sigma \in \Sigma$ and τ a face of σ implies that $\tau \in \Sigma$,
   b) $\sigma \in \Sigma$ and $\tau \in \Sigma$ implies that $\sigma \cap \tau$ is a face of both σ and τ.

7. Simplices of Σ of dimensions 0 are called the vertices of Σ.

The following remark clarifies the relation of abstract and geometric simplicial complexes. Note that when we first encountered this correspondence in Section 1.3, instead $(v_0, \ldots, v_i)$ we used conv$\{v_0, \ldots, v_i\}$ to denote the geometric simplex or equivalently the convex hull. But in order to be able to relate simplicial to cellular homology we have to order the vertices of a geometric simplex and therefore subsequently use the notation $(v_0, \ldots, v_i)$. We refer the reader to the book [37] or to Chapter 1 (around Proposition 1.3.4) for more facts.

Remark 2.3.5. An *abstract simplicial complex* over ground set Ω is a subset $\Delta \subset 2^\Omega$ of the power set of a given (finite) set V such that for all $A \in \Delta$ and all $B \subset A$ we have $B \in \Delta$. In the theory of Combinatorial Algebraic Topology the following facts (see [37]) are well known:

1. For every abstract simplicial complex Δ there exists a geometric simplicial complex Σ and an isomorphism

$$\Phi : \Delta \xrightarrow{\cong} \Sigma$$

   of partially ordered sets, that is, for all $A, B \in \Delta$ such that $A \subset B$ we have $\Phi(A) \subset \Phi(B)$. In this situation we say Σ is a geometric realization of Δ.

2. Two geometric simplicial complexes realizing the same abstract simplicial complex are homeomorphic.

Justified by the previous remark, in the remainder of these lecture notes we will not distinguish between abstract and geometric simplicial complexes.

To see that simplicial complexes can be viewed as a special class of CW-complexes consider the following:

Proposition 2.3.1 (see [11], p.245-247) *Let Σ be a (geometric) simplicial complex. Then there exist characteristic maps $f_\sigma : B^i \longrightarrow X := \bigcup_{\sigma \in \Sigma} \sigma$ that turn X into a CW-complex. The closed cells of this CW-complex are the affine simplices of Σ. One can choose the characteristic maps f_σ such that the coefficients $[\tau : \sigma]$ of the cellular homology of X are given in the following way: If $\sigma = (v_0, \ldots, v_i)$, $\tau = (v_0, \ldots, v_{\nu-1}, v_{\nu+1}, \ldots, v_i)$ then $[\tau : \sigma] = (-1)^i$.*

Indeed simplicial complexes are examples of regular CW-complexes. This concept will be defined and serve as the starting point for discrete Morse theory in Section 4.1 (see also Exercise 2.3.8).

Exercises

Exercise 2.3.6. Let C_i be a free Abelian group $i \geq -1$, where $C_{-1} = 0$. Let $\partial_i : C_i \to C_{i-1}$, $i \geq 0$, be homomorphisms of Abelian groups such that $\partial_i \circ \partial_{i+1} = 0$ for $i \geq 0$. If $C_0/\mathrm{Im}\partial_1 = \mathrm{Ker}\partial_0/\mathrm{Im}\partial_1$ is a free Abelian group. Then there is a CW-complex X such that $(C_i, \partial_i)_{i \geq 0}$ is the cellular chain complex of X.

Exercise 2.3.7. Show that the maps ∂_i, $i \geq 0$, set up in Definition 2.3.1 are indeed differentials; i.e. $\partial_i \circ \partial_{i+1} = 0$.

Exercise 2.3.8. Show that the differential of the chain complex calculating cellular homology of a (geometric) simplicial complex has only coefficients $0, \pm 1$.

Exercise 2.3.9. Let $X_{\mathcal{A}}$ be the CW-complex from Exercise 2.2.7. Show that, the differential of the chain complex calculating cellular homology of $X_{\mathcal{A}}$ has only coefficients $0, \pm 1$.

Exercise 2.3.10. Prove Proposition 2.3.1.

Exercise 2.3.11. Let M be a matroid considered a an abstract simplicial complex. Show that the homology of M is concentrated in $\dim(M)$; that is $\widetilde{H}_i(M; \mathbb{Z}) = 0$ for $i \neq \dim(M)$.

2.4 Cellular Chain Complexes and Cellular Resolutions

In order to link CW-complexes to $\mathbb{Z}^d$-graded modules we need to equip the CW-complex itself with a suitable grading. We define the class of gradings that will become relevant later in this section.

Definition 2.4.1 (Graded CW-complex).

1. For a (not necessarily finite) partially ordered set $(P, \preceq)$ and a map $f : (X, A)^{(*)} \to P$ we call (X, A, f) a P-graded CW-pair if f is order preserving. (See Definition 2.2.2 for the definition of the partial order on $(X, A)^{(*)}$.)

2. If $(P, \preceq)$ is a partially ordered set, (X, f) a P-graded CW-complex and $p \in P$, we denote by $X_{\preceq p}$ the P-graded sub-CW-complex of X consisting of all cells $\sigma \in X^{(*)}$ such that $f(\sigma) \preceq p$.

Example 2.4.2. 1. Let X be a CW-complex and assume that P is a lattice (i.e.; a partially ordered set for which infima and suprema exist). Then any map $f : X^{(0)} \to P$ induces a grading (X, f) on X. Namely, we grade the cell σ be the supremum of all the gradings of the 0-cells in the boundary of σ.

2. Let X be a CW-complex and let P be the set of cells of X ordered by inclusion. Then if f is the map sending the cells of X to their copy in P the pair (X, f) is a graded CW-complex.

$\mathbb{Z}^d$-graded free chain complexes and in particular free resolutions can be constructed from graded CW-complexes in the following way:

Definition 2.4.3 (Cellular chain complex, cellular resolution). Let R be a ring. We call a $\mathbb{Z}^d$-graded free chain complex

$$\mathcal{C} : \ldots \xrightarrow{\partial_{i+1}} \mathcal{C}_i \xrightarrow{\partial_i} \mathcal{C}_{i-1} \xrightarrow{\partial_{i-1}} \ldots \xrightarrow{\partial_1} \mathcal{C}_0, \ \mathcal{C}_i = \bigoplus_{\alpha \in \mathbb{Z}^d} C_i^\alpha, \ C_i^\alpha = R(-\alpha)^{\beta_i^\alpha}$$

cellular if there is a $(\mathbb{Z}^d, \Lambda)$-graded CW-pair (X, A, gr) such that:

1. For all $i \in \mathbb{N}$ there is a basis e_σ of C_i^α indexed by the i-cells σ in $(X, A)^{(i)}$ such that $\mathrm{gr}(\sigma) = \alpha$.

2. For all $i \in \mathbb{N} - \{0\}$, $\sigma \in (X, A)^{(i)}$ we have

$$\partial_i e_\sigma = \sum_{\sigma \geq \sigma' \in X^{(i-1)}} [\sigma' : \sigma] \, \mathbf{x}^{\mathrm{gr}(\sigma) - \mathrm{gr}(\sigma')} \, e_{\sigma'},$$

where $[\sigma' : \sigma]$ is the coefficient of σ' in the differential of σ in the cellular chain complex of the pair (X, A).

In this situation we write $\mathcal{C}_{(X,A)}^{\mathrm{gr}}$ for the given chain complex and say that $\mathcal{C}_{(X,A)}^{\mathrm{gr}}$ is *supported* by the $(\mathbb{Z}^d, \Lambda)$-graded CW-pair (X, A, gr). When the $\mathbb{Z}^d$-graded free chain complex $\mathcal{C}$ is a $\mathbb{Z}^d$-graded free resolution $\mathcal{C} = \mathcal{F}$, we say that $\mathcal{F} = \mathcal{F}_{(X,A)}^{\mathrm{gr}}$ is a $\mathbb{Z}^d$-graded free *cellular resolution supported* by the $(\mathbb{Z}^d, \Lambda)$-graded CW-pair (X, A, gr).

Example 2.4.1 *Consider the ideal* $M = (x_1^2 x_2, x_2^2 x_3, x_1 x_3^2, x_1 x_2 x_3)$ *in* $S = k[x_1, x_2, x_3]$. *The multigraded minimal free resolution of* M *over* S *is given by:*

$$\mathcal{F}: 0 \longrightarrow \begin{matrix} S(-(2,1,1)) \\ \oplus \\ S(-(1,2,1)) \\ \oplus \\ S(-(1,1,2)) \end{matrix} \xrightarrow{\begin{pmatrix} x_3 & 0 & 0 \\ 0 & x_1 & 0 \\ 0 & 0 & x_2 \\ -x_1 & -x_2 & -x_3 \end{pmatrix}} \begin{matrix} S(-(2,1,0)) \\ \oplus \\ S(-(0,2,1)) \\ \oplus \\ S(-(1,0,2)) \\ \oplus \\ S(-(1,1,1)) \end{matrix} \longrightarrow M \longrightarrow 0$$

This resolution is easily seen to be cellular with $\mathcal{F} = \mathcal{F}_X^{\mathrm{gr}}$ *for the graded complex* (X, gr) *from Fig. 2.2:*

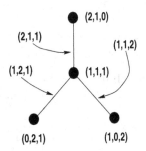

Fig. 2.2. Graded CW-complex from Example 2.4.4

The above example is a special case of Example 2.4.2 (1) for the given CW-complex and the lattice $\mathbb{N}^3$ – *note that here we use lattice in the order theoretic sense* – *with partial order given by componentwise comparison in* $\mathbb{N}$.

Exercises and Questions

Exercise 2.4.4. Show that the resolutions from Exercise 2.1.16 are cellular and describe the corresponding complexes.

Exercise 2.4.5. Assume the $\mathbb{Z}^d$-graded $k[x_1, \ldots, x_d]$-modules M and M' have cellular minimal free resolutions. Show that $M \oplus M'$ also has a cellular minimal free resolution.

Question 2.4.6. Assume there are cellular minimal free resolutions $\mathcal{F}$ and $\mathcal{F}'$ of a $\mathbb{Z}^d$-graded $k[x_1, \ldots, x_d]$-module M. Let X be the CW-complex which supports $\mathcal{F}$ and let X' be the CW-complex that supports $\mathcal{F}'$. Are X and X' homeomorphic ? homotopy equivalent ?

2.5 Co-Artinian Monomial Modules

In this section we introduce the concept of a co-Artinian monomial module. This notion was coined in [6] where the authors also exhibit cellular resolutions for them. Monomial modules will be one of the central examples that demonstrate the use of discrete Morse theory for the construction of minimal free resolutions in the next chapter. It does not immediately connect to the concepts defined so far in this chapter, but all ingredients will come together in Chapter 3.3.

Monomial modules generalize the concept of a monomial ideal and also capture modules whose structure is determined by a sublattice of the integer lattice. In the following we denote by

$$T := k[x_1^{\pm 1}, \ldots, x_d^{\pm 1}] = \bigoplus_{\alpha \in \mathbb{Z}^d} k\mathbf{x}^\alpha$$

the ring of Laurent polynomials generated by all monomials $\mathbf{x}^\alpha = x_1^{\alpha_1} \cdot \ldots \cdot x_d^{\alpha_d}$ for $\alpha = (\alpha_1, \ldots, \alpha_d) \in \mathbb{Z}^d$.

Definition 2.5.1. *1. An $S = k[x_1, \ldots, x_d]$-module M is called monomial module if it is a submodule of T generated by monomials $\mathbf{x}^\alpha$ for $\alpha \in \mathbb{Z}^d$.*
2. A monomial module M is called co-Artinian if for each $\mathbf{x}^\alpha \in M$ there are only finitely many $\mathbf{x}^\beta \in M$ such that $\mathbf{x}^{\alpha - \beta} \in S$.
3. Let M be a monomial module. A monomial $\mathbf{x}^\alpha \in M$ is called minimal if $\frac{\mathbf{x}^\alpha}{x_i} \notin M$ for all $i = 1, \ldots, d$. We denote by $\mathrm{MinGen}(M)$ the set of minimal monomials in M.

The following lemma gives an alternative definition of co-Artinian monomial modules. It shows why the conditions from Definition 2.5.1 are well suited for our combinatorial setting in which all modules in a resolution are finitely generated.

Lemma 2.5.2 ([6]). *A monomial module M is co-Artinian if and only if it is generated by the set $\mathrm{MinGen}(M)$ of its minimal elements.*

Example 2.5.3. An ideal in $S = k[x_1, \ldots, x_d]$ that is generated by monomials is called a *monomial ideal*. Of course, any monomial ideal is an example of a co-Artinian monomial module.

We close by showing that the class of co-Artinian monomial modules is substantially richer than the class of monomial ideals.

Definition 2.5.4. *A subgroup $L \subset \mathbb{Z}^d$ of the additive group $\mathbb{Z}^d$ is called integer lattice .*

The following obvious lemma provides another class of examples of co-Artinian monomial modules:

Lemma 2.5.5 ([6]). *Let $L \subset \mathbb{Z}^d$ be an integer lattice. Let $M_L \subset T$ be the monomial module generated by all monomials $\mathbf{x}^\alpha$ such that $\alpha \in L$. Then M_L is co-Artinian if and only if $L \cap \mathbb{N}^d = \{(0, \ldots, 0)\}$.*

Example 2.5.6. If $\Lambda = \mathbb{N}\lambda_1 + \cdots + \mathbb{N}\lambda_r \subseteq \mathbb{N}^d$ is an affine semigroup. Then the lattice $L_\Lambda = \mathbb{Z}\lambda_1 + \cdots + \mathbb{Z}\lambda_r$ generated by Λ is co-Artinian if and only if $\Lambda = \{(0, \ldots, 0)\}$.

Exercises

Exercise 2.5.7. Show that for a monomial ideal there is a unique minimal set of monomials that generate the ideal.

Exercise 2.5.8. Let I be a monomial ideal in $k[x_1, \ldots, x_d]$. Assume that I is minimally generated by the monomials $m_1, \ldots, m_f$. Let s_i be the maximal exponent of the variable x_i in the monomials $m_1, \ldots, m_f$. Consider the polynomial ring $k[x_i^{(j)} \mid 1 \le i \le d, 1 \le j \le s_i]$. To each monomial m_j we associate the monomial m_j^{pol} where we replace the maximal power x_i^l of x_i in m_j by $x_i^{(1)} \cdots x_i^{(l)}$. Now the ideal I^{pol} generated by the monomials m_j^{pol}, $1 \le j \le f$, is called the polarization of I.

1. Given a minimal free resolution of I, construct a minimal free resolution of I^{pol}.

2. Use (1) to show that the Betti numbers β_i of I^{pol} and I coincide.

3

Cellular Resolution

This chapter contains more facts about cellular resolutions and in particular many examples of cellular resolutions. The set of these examples is chosen with some personal bias from a big set of examples of cellular resolutions that have emerged over the last years. We try to be a bit more complete by covering in the exercises some of the examples that are left out.

3.1 When Does a CW-Complex Support a Cellular Resolution ?

In order to formulate a criterion for a CW-complex to support a give $\mathbb{Z}^d$-graded chain complex $\mathcal{C}$ we need to define certain subcomplexes of $\mathcal{C}$.

Definition 3.1.1. *Let*

$$\mathcal{C} \; : \; \ldots \xrightarrow{\partial_{i+1}} \mathcal{C}_i \xrightarrow{\partial_i} \mathcal{C}_{i-1} \xrightarrow{\partial_{i-1}} \ldots \xrightarrow{\partial_1} \mathcal{C}_0,$$

$\mathcal{C}_i = \bigoplus_{\alpha \in \mathbb{Z}^d} C_i^\alpha$, $C_i^\alpha = R(-\alpha)^{\beta_i^\alpha}$ *be a $\mathbb{Z}^d$-graded free chain complex over the affine semigroup ring $R = k[\Lambda]$. Recall, that Λ induces a partial order on $\mathbb{Z}^d$ by $\alpha \preceq \beta$ if and only if $\alpha + \gamma = \beta$ for some $\gamma \in \Lambda$.*

1. *We denote by $\mathcal{C}^{\preceq \alpha}$ ($\mathcal{C}^{\prec \alpha}$ respectively) the subsequence of $\mathcal{C}$ consisting of the submodules $C_i^{\preceq \alpha} := \bigoplus_{\beta \preceq \alpha} C_i^\beta$ ($C_i^{\prec \alpha} := \bigoplus_{\beta \prec \alpha} C_i^\beta$ respectively). These are again $\mathbb{Z}^d$-graded free chain complexes. They are the sub-chain complexes of $\mathcal{C}$ generated by all generators with multidegree $\preceq \alpha$ (respectively $\prec \alpha$).*
2. *Consider the decomposition $\mathcal{C}_i = \bigoplus_{\alpha \in \mathbb{Z}^d} (\mathcal{C}_i)_\alpha$ of $\mathcal{C}_i$ as a $k-$module. That is, for all $\alpha \in \mathbb{Z}^d$, $(\mathcal{C}_i)_\alpha$ consists of all elements of $\mathcal{C}_i$ of multidegree α. By $\mathcal{C}_\alpha$ we denote the chain complex of k-vectorspaces given by the degree $\alpha \in \mathbb{Z}^d$ components $(\mathcal{C}_i)_\alpha$. We call $\mathcal{C}_\alpha$ the degree α strand of $\mathcal{C}$.*

Note that $(\mathcal{C}_i)_\alpha \neq \mathcal{C}_i^\alpha$. In general neither one is a subset of the other.

Example 3.1.2. Let $S = k[x_1, \ldots, x_n]$ and $e_i, 1 \leq i \leq d$, the i-th unit vector. Set $C_i = \bigoplus\limits_{A \subseteq [d]; \# A = i} S(-\sum\limits_{i \in A} e_i)$. Let us choose basis element e_A for each $S(-\sum\limits_{i \in A} e_i)$. If $A = \{j_1 < \cdots < j_i\}$ then we define

$$\partial_i(e_A) = \sum_{l=1}^{i}(-1)^l x_{j_l} e_{A \backslash j_l}.$$

Then

$$\mathcal{C}: 0 \xrightarrow{\partial_{d+1}} C_d \xrightarrow{\partial_d} \cdots \to C_0$$

is a free chain complex – indeed a minimal free resolution of the field k over S. Now if we choose $\alpha = (1, \ldots, 1)$ then $\mathcal{C}^{\preceq \alpha} = \mathcal{C}$ and

$$\mathcal{C}^{\prec \alpha}: 0 \to C_{d-1} \xrightarrow{\partial_{d-1}} \to \cdots \xrightarrow{\partial_1} C_0.$$

The complex of k-vector spaces $\mathcal{C}_\alpha$ is given by

$$\mathcal{C}_\alpha: 0 \to k \to k^{\binom{d}{1}} \to k^{\binom{d}{2}} \to \cdots \to k^{\binom{d}{d}},$$

with trivial differential.

Remark 3.1.1 *Since the differentials ∂_i in Definition 3.1.1 are homogeneous and R−linear, they restrict to differentials of $\mathcal{C}^{\prec \alpha}$, $\mathcal{C}^{\preceq \alpha}$ and $\mathcal{C}_\alpha$, so the above chain complexes are well-defined.*

It is helpful to establish the following criterion for a $(\mathbb{Z}^d, \Lambda)$-graded pair (X, A, gr) to support a cellular resolution.

Proposition 3.1.2 ([6]) *Let $R = k[\Lambda]$ be an affine semigroup ring.*

1. *The $(\mathbb{Z}^d, \Lambda)$-graded CW-pair (X, A, gr) supports a cellular resolution of a $\mathbb{Z}^d$-graded R-module if and only if for all $\alpha \in \mathbb{Z}^d$ we have that $H_i(X_{\preceq \alpha}, A_{\preceq \alpha}; k) = 0$ for all $i \geq 1$.*
2. *The $(\mathbb{Z}^d, \Lambda)$-graded CW-complex (X, gr) supports a cellular resolution of a monomial R-module if and only if for all $\alpha \in \mathbb{Z}^d$ the subcomplex $X_{\preceq \alpha}$ is acyclic over k or empty.*
3. *A cellular resolution supported by a $(\mathbb{Z}^d, \Lambda)$-graded pair (X, A, gr) is minimal if and only if for all cells $\sigma, \sigma' \in (X, A)^{(*)}$, $\sigma' \leq \sigma$ and $\dim \sigma' = \dim \sigma - 1$ we have either $\mathrm{gr}(\sigma) \neq \mathrm{gr}(\sigma')$ or $[\sigma' : \sigma] = 0$.*

Proof. For any multidegree α the degree α strand $(\mathcal{F}_{X,A})_\alpha$ is supported by basis elements e_σ for cells σ in $(X_{\preceq\alpha}, A_{\preceq\alpha})^{(*)}$. Let $z := \sum_{\sigma \in (X_{\preceq\alpha}, A_{\preceq\alpha})^i} a_\sigma \cdot$ $\underline{\mathbf{x}}^{\alpha - \mathrm{gr}\sigma} \cdot e_\sigma$. Then $\partial_i z = 0$ if and only if the coefficient $b_{\sigma'}$ of any $e_{\sigma'}$, $\sigma' \in$ $(X_{\preceq\alpha}, A_{\preceq\alpha})^{(i-1)}$, in $\partial_i z$ is zero. Now $b_{\sigma'} = \underline{\mathbf{x}}^{\alpha - \mathrm{gr}\sigma'} \sum_{\sigma \in (X_{\preceq\alpha}, A_{\preceq\alpha})^{(i)}} a_\sigma \cdot [\sigma' : \sigma]$.

Thus $b_{\sigma'} = 0$ if and only if $\sum_{\sigma \in (X_{\preceq\alpha}, A_{\preceq\alpha})^{(i)}} a_\sigma \cdot [\sigma' : \sigma] = 0$. But this is equivalent to the fact that $\sum_{\sigma \in (X_{\preceq\alpha}, A_{\preceq\alpha})^i} a_\sigma \cdot \sigma$ is a cycle in the cellular chain complex of $(X_{\preceq\alpha}, A_{\preceq\alpha})$ with coefficients in k.

Similar arguments show that $z := \sum_{\sigma \in (X_{\preceq\alpha}, A_{\preceq\alpha})^{(i)}} a_\sigma \cdot \underline{\mathbf{x}}^{\alpha - \mathrm{gr}\sigma} \cdot e_\sigma$ is a boundary $z = \partial_{i+1} z'$ if and only if $\sum_{\sigma \in (X_{\preceq\alpha}, A_{\preceq\alpha})^{(i)}} a_\sigma \cdot \sigma$ is a boundary in the cellular chain complex of $(X_{\preceq\alpha}, A_{\preceq\alpha})$ with coefficients in k.

It follows from Definition 2.1.11 that a resolution $\mathcal{F}_{X,A}$ is minimal if and only if no entry in the matrix describing the differential is a non-zero constant. For cellular resolutions this condition directly translates into the last assertion of the proposition.

Example 3.1.3. We adopt the notation from Example 3.1.2. Let $\Delta = 2^{[d]}$ be the full simplex on ground set $[d]$. If we grade the simplex $A \in \Delta$ by $\mathrm{gr}(A) = \sum_{i \in A} e_i$, then it is easily seen that the resolution $\mathcal{C}$ is supported by (Δ, gr). Now for any degree $\alpha \in \mathbb{Z}^d$ we have that $\Delta_{\preceq\alpha}$ is either empty or 2^A for $A = \mathrm{supp}\alpha = \{i \in [d] \mid \alpha_i \neq 0\}$. Thus indeed (Δ, gr) fulfills the criterion from Proposition 3.1.2 (2).

We would like to mention that graded CW-complexes are of interest even if they do not support a free resolution. We refer the reader to recent work by Floystad [19] for more details and interesting applications to Cohen-Macaulay simplicial complexes.

Questions and Exercises

Exercise 3.1.4. Show that there is a monomial ideal in some $k[x_1, \dots, x_d]$ for which the minimal free resolution is cellular for all k but there is no CW-complex X that supports the minimal free resolution for all k.

Exercise 3.1.5. Show that there is a monomial ideal whose minimal free resolution cannot be supported by a regular CW-complex. (This is a hard exercise, see [47] for a counterexample and [60] for structural insight into the class of counterexamples.)

Question 3.1.6. Velasco [60] showed that there are monomial ideals for which the minimal free resolution cannot be supported by a CW-complex. What can be said about the class of monomial ideals for which there is a cellular minimal free resolution ? Does this class have nice algebraic properties ? What can be said about the class of simplicial complexes for which the Stanley-Reisner ideal has a cellular minimal free resolution ?

3.2 Reading off the Betti Numbers

By the following lemma the (multigraded) Betti numbers β_i^α, $\alpha \in \mathbb{Z}^d$, of the minimal $\mathbb{Z}^d$-graded free resolution of M can be computed from any given $\mathbb{Z}^d$-graded free resolution $\mathcal{F}$ of M:

Lemma 3.2.1 *Let M be a $\mathbb{Z}^d$-graded R-module, $\mathcal{F}$ a $\mathbb{Z}^d$-graded free resolution of M. Then:*

$$\beta_i^\alpha(M) = \dim H_i((\mathcal{F}^{\preceq \alpha})_\alpha, (\mathcal{F}^{\prec \alpha})_\alpha).$$

Proof. If $\mathcal{F} = \mathcal{F}_{\min}$ is minimal, the assertion follows from the fact that all differentials of the chain complex $((\mathcal{F}^{\preceq \alpha})_\alpha, (\mathcal{F}^{\prec \alpha})_\alpha)$ vanish. For general $\mathcal{F}$ there is a direct sum decomposition $\mathcal{F} = \mathcal{F}_{\min} \oplus \mathcal{R}$, where $\mathcal{R}$ is a direct sum of exact sequences

$$\mathcal{R}_{\alpha,\gamma,i} : \cdots \to 0 \to S(-\alpha)^\gamma \to S(-\alpha)^\gamma \overset{i \ times}{\overbrace{\to 0}} \to 0,$$

$\alpha \in \mathbb{Z}^d$, $i, \gamma \in \mathbb{N}$. This gives:

$$H_i((\mathcal{F}^{\preceq \alpha})_\alpha, (\mathcal{F}^{\prec \alpha})_\alpha) \cong H_i((\mathcal{F}_{\min}^{\preceq \alpha} \oplus \mathcal{R}^{\preceq \alpha})_\alpha, (\mathcal{F}_{\min}^{\prec \alpha} \oplus \mathcal{R}^{\prec \alpha})_\alpha)$$
$$\cong H_i((\mathcal{F}_{\min}^{\preceq \alpha})_\alpha, (\mathcal{F}_{\min}^{\prec \alpha})_\alpha) \oplus H_i((\mathcal{R}^{\preceq \alpha})_\alpha, (\mathcal{R}^{\prec \alpha})_\alpha)$$

and the assertion follows because the sequence $((\mathcal{R}^{\preceq \alpha})_\alpha, (\mathcal{R}^{\prec \alpha})_\alpha)$ contributes no homology.

Starting with a cellular resolution $\mathcal{F}_{X,A}$ of a $\mathbb{Z}^d$-graded R-module M and following the idea of Lemma 3.2.1 it is possible to describe the Betti numbers of the minimal free $\mathbb{Z}^d$-graded resolution of M in terms of the homology of the underlying $(\mathbb{Z}^d, \Lambda)$-graded CW-pair (X, A, gr). We write $H_i(X, A; k)$ for the i-th cellular homology group of the pair (X, A) with coefficients in k and $\widetilde{H}_i(X; k)$ for the reduced homology of X with coefficients in k. As usual $H_i(\emptyset; k) = 0$ for $i \neq -1$ and $H_{-1}(\emptyset; k) = k$.

Proposition 3.2.2 *Let $R = k[\Lambda]$ be an affine semigroup ring. Let M be a $\mathbb{Z}^d$-graded R-module and (X, A, f) a $\mathbb{Z}^d$-graded CW-pair supporting a $\mathbb{Z}^d$-graded free resolution of M. Then:*

$$\beta_i^\alpha(M) = \dim H_i(X_{\preceq\alpha}, X_{\prec\alpha} \cup A_{\preceq\alpha}; k)$$

$$= \begin{cases} \dim H_{i-1}(X_{\prec\alpha}, A_{\prec\alpha}; k) & \text{if } A_{\preceq\alpha} \neq \emptyset \\ \dim \tilde{H}_{i-1}(X_{\prec\alpha}; k) & \text{if } A_{\preceq\alpha} = \emptyset \end{cases},$$

where the first equation holds for all $i \geq 0$, the second for all $i \geq 2$. If in addition $X_{\prec\alpha}$ and $A_{\prec\alpha}$ are acyclic or empty, the second equation also holds for $i = 1$. Note that for $i = 0$ the righthand side of the first equation is easy to calculate.

In particular, $\beta_i^\alpha = 0$ in case there is no i-cell in $(X, A)^{(i)}$ of degree α.

Proof. Denoting the i–th cellular chain group of a CW-pair by $C_i(X, A)$, the chain-complex $((\mathcal{F}_{X,A}^{\preceq\alpha})_\alpha, (\mathcal{F}_{X,A}^{\prec\alpha})_\alpha)$ reads as follows:

$$\cdots \to \frac{C_{i+1}(X_{\preceq\alpha}, A_{\preceq\alpha})}{C_{i+1}(X_{\prec\alpha}, A_{\prec\alpha})} \to \frac{C_i(X_{\preceq\alpha}, A_{\preceq\alpha})}{C_i(X_{\prec\alpha}, A_{\prec\alpha})} \to \frac{C_{i-1}(X_{\preceq\alpha}, A_{\preceq\alpha})}{C_{i-1}(X_{\prec\alpha}, A_{\prec\alpha})} \to \cdots,$$

with induced differentials. Since we have:

$$\frac{C_i(X_{\preceq\alpha}, A_{\preceq\alpha})}{C_i(X_{\prec\alpha}, A_{\prec\alpha})} \cong \frac{C_i(X_{\preceq\alpha})/C_i(A_{\preceq\alpha})}{C_i(X_{\prec\alpha})/C_i(A_{\prec\alpha})} \cong \qquad (3.2.1)$$

$$\cong \frac{C_i(X_{\preceq\alpha})}{C_i(X_{\prec\alpha}) + C_i(A_{\preceq\alpha})} \cong \frac{C_i(X_{\preceq\alpha})}{C_i(X_{\prec\alpha} \cup A_{\preceq\alpha})} \cong C_i(X_{\preceq\alpha}, X_{\prec\alpha} \cup A_{\preceq\alpha})$$

and the induced differentials are the same as those of the cellular chain complex of the pair $(X_{\preceq\alpha}, X_{\prec\alpha} \cup A_{\preceq\alpha})$, applying Lemma 3.2.1 proves the first equation.

Applying equation (3.2.1), the long exact sequence of the pair $(C_\bullet(X_{\prec\alpha}, A_{\prec\alpha}), C_\bullet(X_{\preceq\alpha}, A_{\preceq\alpha}))$ can be written as follows:

$$\cdots \to H_{i+1}(X_{\preceq\alpha}, A_{\preceq\alpha}) \to H_{i+1}(X_{\preceq\alpha}, X_{\prec\alpha} \cup A_{\preceq\alpha}) \longrightarrow$$
$$\to H_i(X_{\prec\alpha}, A_{\prec\alpha}) \to H_i(X_{\preceq\alpha}, A_{\preceq\alpha}) \to \cdots$$

This proves the second equation for $i \geq 2$.

Assume now $X_{\prec\alpha}$ and $A_{\prec\alpha}$ to be acyclic or empty. The long exact sequence for pairs yields the exact sequence:

$$0 = \tilde{H}_1(X_{\preceq\alpha}; k) \to \tilde{H}_1(X_{\preceq\alpha}, X_{\prec\alpha} \cup A_{\preceq\alpha}; k) \to$$
$$\to \tilde{H}_0(X_{\prec\alpha} \cup A_{\preceq\alpha}; k) \to \tilde{H}_0(X_{\preceq\alpha}; k) = 0$$

If $A_{\preceq \alpha} = \emptyset$, we have $\tilde{H}_1(X_{\preceq \alpha}, X_{\prec \alpha} \cup A_{\preceq \alpha}; k) = \tilde{H}_0(X_{\prec \alpha}; k)$ as stated.

If $A_{\preceq \alpha}$ is acyclic, considering that $X_{\prec \alpha} \cap A_{\preceq \alpha} = A_{\prec \alpha}$, the Mayer-Vietoris-sequence for $X_{\prec \alpha} \cup A_{\preceq \alpha}$ yields $\tilde{H}_0(A_{\prec \alpha}; k) \to \tilde{H}_0(X_{\prec \alpha}; k) \to \tilde{H}_0(X_{\prec \alpha} \cup A_{\preceq \alpha}; k) \to \tilde{H}_{-1}(A_{\prec \alpha}; k) \to \tilde{H}_{-1}(X_{\prec \alpha}; k)$, while the long exact sequence for the pair $(A_{\prec \alpha}, X_{\prec \alpha})$ yields $\tilde{H}_0(A_{\prec \alpha}; k) \to \tilde{H}_0(X_{\prec \alpha}; k) \to H_0(X_{\prec \alpha}, A_{\prec \alpha}; k) \to \tilde{H}_{-1}(A_{\prec \alpha}; k) \to \tilde{H}_{-1}(X_{\prec \alpha}; k)$. The assertion follows from the five lemma.

If there is no i-cell of degree α in $(X, A)^{(i)}$ then $H_i(X_{\preceq \alpha}, X_{\prec \alpha} \cup A_{\preceq \alpha}; k) = H_i(X_{\preceq \alpha}, X_{\preceq \alpha}; k) = 0$. This proves that $\beta_i^\alpha = 0$ in case there is no i-cell of degree α in $(X, A)^{(i)}$.

Before we close this section we want to study the relative situation; that is CW-Pairs and quotients of modules.

In a situation where we have a $\mathbb{Z}^d$-graded free resolution $\mathcal{F}$ of a $\mathbb{Z}^d$-graded module M and an exact subsequence $\mathcal{G} \hookrightarrow \mathcal{F}$ resolving the submodule $N \hookrightarrow M$, passing to quotients yields a $\mathbb{Z}^d$-graded free resolution of the quotient M/N if and only if the modules $\mathcal{G}_i$ are direct summands of $\mathcal{F}_i$ for all $i \geq 0$. The following cellular situation fulfills this condition:

Proposition 3.2.3 *Let the CW-pair (Y, A) support a $\mathbb{Z}^d$-graded cellular free resolution of the $\mathbb{Z}^d$-graded S-module M which restricts to a $\mathbb{Z}^d$-graded cellular free resolution of a submodule $N \hookrightarrow M$ supported by the CW-subpair (X, A) of (Y, A). Passing to quotients yields a $\mathbb{Z}^d$-graded cellular free resolution of M/N supported on the CW-pair (Y, X).*

The proof of Proposition 3.2.3 is straightforward and left to the reader (see Exercise 3.2.2).

Now that we have gathered sufficient theory around cellular resolutions we are need of classes of examples. These will be provided the following sections.

Exercises

Exercise 3.2.1. Let $\Omega_{d,l}$ be the set of squarefree monomials of degree $l + 1$ in $S = k[x_1, \ldots, x_d]$. Let $\Delta_{d,l}$ be the full simplex $2^{\Omega_{d,l}}$. We grade a simplex $A \in \Delta_{d,l}$ be the least common multiple of the monomials in A.

1. Show that $\Delta_{d,l}$ supports a resolution of $I_{U_{d,l}}$. This is a special case of the Taylor resolution which will be defined as the first example of a cellular resolution at the beginning of Section 3.3.
2. Read off the Betti numbers of $I_{U_{d,l}}$ from this resolution.

Exercise 3.2.2. Prove Proposition 3.2.3.

3.3 Examples of Cellular Resolutions

In the following sections we present several classes of cellular resolutions. Even though the Taylor resolution has been known for about 40 years it has only after [6] been seen as a geometric construction. Similarly, the Bar resolution is a classical object from Homological Algebra, and clearly had its geometric flavor in terms of simplicial set. On the other hand the explicit use of its cellular structure in order to find minimal free resolution seems to be rather recent. More results and details about the LCM- and LCM*-resolution can also be found in [4].

The Taylor Resolution:

Recall that for a co-Artinian monomial module M we denote by $\mathrm{MinGen}(M)$ its uniquely defined minimal set of monomial generators (See Definition 2.5.1). In the sequel we will set up a cellular resolution for co-Artinian monomial modules M and quotients, which in general is very far from being minimal. This construction in case of monomial ideals dates back to the 1960's and can be found in the PhD thesis of Diana Taylor [59] – even though it was not seen as a cellular resolution at that time.

Definition 3.3.1 (Taylor complex)   *1. For a co-Artinian monomial module M let $\mathfrak{T}(M)$ be the simplicial complex on the ground set $\mathrm{MinGen}(M)$ and simplices the finite subsets of $\mathrm{MinGen}(M)$. We grade the simplices $\sigma \in \mathfrak{T}(M)$ by $\mathrm{lcm}(\sigma) := \mathrm{lcm}\{m | m \in \sigma\}$. Via the correspondence $\alpha \leftrightarrow \underline{\mathbf{x}}^{\alpha}$, we also regard this grading as a $\mathbb{Z}^d$–grading. $\mathfrak{T}(M)$ is called the Taylor complex of M.*

2. More generally, let $N \subset M$ be two co-Artinian monomial modules. Then we denote by $\mathfrak{T}(M, N)$ the simplicial complex on the ground set $\mathrm{MinGen}(M) \cup \mathrm{MinGen}(N)$ and simplices the finite subsets of $\mathrm{MinGen}(M) \cup \mathrm{MinGen}(N)$. Again, we grade the simplices $\sigma \in \mathfrak{T}(M, N)$ by $\mathrm{lcm}(\sigma) := \mathrm{lcm}\{m | m \in \sigma\}$ and call $\mathfrak{T}(M, N)$ the Taylor complex of the pair (M, N).

Lemma 3.3.2 *Let M and N be two co-Artinian monomial modules such that $N \subset M$. Then $\mathfrak{T}(M, N)$ and $\mathfrak{T}(M)$ define cellular resolutions of the quotient module M/N and of the monomial module M. module.*

Proof. By Proposition 3.1.2 it suffices to show that $\mathfrak{T}(M, N)_{\leq \alpha}$ and $\mathfrak{T}(M)_{\leq \alpha}$ are acyclic or empty complexes for all $\alpha \in \mathbb{Z}^d$. If $\mathfrak{T}(M, N)_{\leq \alpha} \neq \emptyset$ or $\mathfrak{T}(M)_{\leq \alpha} \neq \emptyset$, then the simplex which contains all points of multidegree α or less is the unique maximal simplex of the complex. Thus $\mathfrak{T}(M, N)_{\leq \alpha}$ or $\mathfrak{T}(M)_{\leq \alpha}$ is contractible and hence acyclic.

Now we are in position to define the resolution due to Diana Taylor [59].

Definition 3.3.3 (Taylor resolution) *1. For a co-Artinian monomial module M the cellular resolution supported by $\mathfrak{T}(M)$ is called the Taylor resolution of M.*

2. Let $N \subset M$ be two co-Artinian monomial modules. The cellular resolution supported by $\mathfrak{T}(M, N)$ is called the Taylor resolution of the pair (M, N).

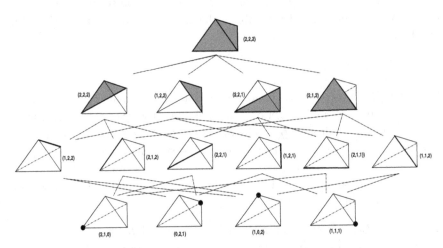

Fig. 3.1. Taylor resolution for Example 2.4.1

Fig. 3.1 provides a geometric picture of the Taylor resolution for the ideal from Example 2.4.4 The following theorem summarizes the facts we have collected so far about the Taylor resolution.

Theorem 3.3.4 *Let $N \subset M$ be co-Artinian monomial modules.*

Then there exists a cellular $\mathbb{Z}^d$-graded free resolution of M/N supported by the pair $(\mathfrak{T}(M, N), \mathfrak{T}(N), \mathrm{lcm})$. In particular, for the Betti numbers $\beta_i^\alpha(M/N)$, $\alpha \in \mathbb{Z}^d$, of the minimal $\mathbb{Z}^d$-graded free resolution of M/N we have for $i \geq 0$:

$$\beta_i^\alpha(M/N) = \begin{cases} \dim H_{i-1}(\mathfrak{T}(M, N)_{\prec\alpha}, \mathfrak{T}(N)_{\prec\alpha}; k) & \text{if } \mathfrak{T}(N)_{\preceq\alpha} \neq \emptyset \\ \dim \tilde{H}_{i-1}(\mathfrak{T}(M, N)_{\prec\alpha}; k) & \text{if } \mathfrak{T}(N)_{\preceq\alpha} = \emptyset \end{cases}.$$

Proof. Follows immediately from Lemma 3.3.2, Proposition 3.2.3 and Proposition 3.2.2.

By the preceding theorem the Taylor resolution of the pair (M, N) indeed is a resolution of the quotient M/N.

Definition 3.3.5 *We call the cellular resolution of M/N supported by the pair $(\mathfrak{T}(M, N), \mathfrak{T}(N), \mathrm{lcm})$ the Taylor resolution of M/N.*

In order to formulate a condition for a monomial module under which the Taylor resolution is minimal we need the following definition and obvious remark:

Definition 3.3.6   *1. Let $p_1, \ldots, p_n \in S = k[x_1, \ldots, x_d]$ be polynomials. The sequence $p_1, \ldots, p_n$ is called a regular sequence (see also Section 1.3) if $\langle p_1, \ldots, p_n \rangle \neq S$ and if for all $i = 1, \ldots, n$ we have that p_i is not a zerodivisor in $S/\langle p_1, \ldots, p_{i-1} \rangle$. Here, $\langle p_1, \ldots, p_{i-1} \rangle \subset S$ denotes the ideal generated by the polynomials $p_1, \ldots, p_{i-1}$.*

2. If $p_1, \ldots, p_n$ is a regular sequence, the ideal $\langle p_1, \ldots, p_n \rangle \subset S$ is called complete intersection .

Remark 3.3.7 *Let $I \subset S$ be a monomial ideal. I is a complete intersection if and only if the elements of $\mathrm{MinGen}(I)$ have disjoint supports, that is, if $m, m' \in \mathrm{MinGen}(I)$, $m \neq m'$ and $x_r \mid m$ for some variable x_r, then $x_r \nmid m'$.*

Proposition 3.3.8   *1. Let $N \subset M$ be co-Artinian monomial modules. The Taylor resolution of M/N is minimal if and only if for all $\sigma \in \mathfrak{T}(M, N)$ and all $m \in \sigma$ we have that $m \mid \mathrm{lcm}(\sigma \backslash \{m\})$ implies $\sigma \backslash \{m\} \in \mathfrak{T}(N)$. In particular the Taylor resolution of M is minimal if and only if for all $\sigma \in \mathfrak{T}(M)$ and all $m \in \sigma$ we have $m \nmid \mathrm{lcm}(\sigma \backslash \{m\})$.*

2. Let $I \subset S = k[x_1, \ldots, x_d]$ be a monomial ideal which is a complete intersection. Then the Taylor resolution of I is minimal.

Proof. Minimality of the Taylor resolution of M/N occurs if and only if for all simplices $\sigma \in \mathfrak{T}(M, N)$ and all $m \in \sigma$ either the coefficient $\mathrm{lcm}\,\sigma/\mathrm{lcm}(\sigma \backslash \{m\})$ is an element of the maximal homogeneous ideal $\mathfrak{m}$ or $\sigma \backslash \{m\} \in \mathfrak{T}(N)$. This proves the first part of the Proposition. The second part is an easy application of the first part.

The Scarf Resolution:

Proposition 3.3.8 shows that the Taylor resolution is minimal only in very exceptional cases. In search of smaller resolutions Proposition 3.3.8 suggests the following strategy: Start with the Taylor resolution and remove those simplices that obstruct minimality, namely the simplices σ with the following property:

$$\text{There exists a monomial } m \in \sigma \text{ such that } m \mid \mathrm{lcm}(\sigma \backslash \{m\}). \qquad (3.3.1)$$

Not very surprisingly minimality is not so easily achieved. The reason is that in general the resulting simplicial complex is not acyclic. Consider the following simple example:

Example 3.3.9 *Let I be the ideal in $k[x_1, x_2]$ generated by the monomials x_1^2, x_2^2 and $x_1 x_2$.*

In Fig. 3.2 we present the graded CW-complex underlying the Taylor resolution in this case.

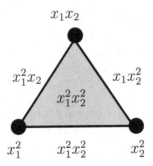

Fig. 3.2. Taylor complex for Example 3.3.9

Here, the simplex $\sigma = \{x_1^2, x_2^2, x_1 x_2\}$ has property (3.3.1) with $m = x_1 x_2$. Thus, above strategy calls for removing σ. But since no other simplex fulfills property (3.3.1) the resulting simplicial complex forms the boundary of a triangle which is not acyclic. In this example this can be fixed by also deleting the simplex $\sigma' = \{x_1^2, x_2^2\}$. The result is a minimal resolution.

Thus, the next idea is not only to remove the simplices $\sigma = \{m_0, \ldots, m_s\}$ with property (3.3.1) but also the corresponding simplices $\sigma = \{m_0, \ldots, m_s\}$ $-\{m\}$. That is we delete all simplices $\sigma = \{m_0, \ldots, m_s\}$ with the property:

There exists $m \in \mathrm{MinGen}(M)$ such that $m \big| \mathrm{lcm}(\sigma \setminus \{m\})$. (3.3.2)

The resulting simplicial complex is called the Scarf complex :

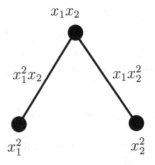

Fig. 3.3. Scarf complex for Example 3.3.9

Definition 3.3.10 (Scarf resolution)   *1. Let $M \subset T = k[x_1^{\pm 1}, \ldots, x_d^{\pm 1}]$ be a co-Artinian monomial module. The simplicial subcomplex of the Taylor complex of M that is given by the set of those simplices σ such that there is no other simplex $\tau \neq \sigma$ with $\mathrm{lcm}\ \sigma = \mathrm{lcm}\ \tau$ is called the Scarf complex of M.*

2. If the Scarf complex of M is acyclic, we call the resolution supported by it, the Scarf resolution of M.

In Fig. 3.3 we depict the Scarf complex for Example 3.3.9. Unfortunately, the Scarf complex in general does not provide a resolution. Consider the next simple example:

Example 3.3.11 *Let I be the ideal in $k[x_1, x_2, x_3]$ generated by the monomials $x_1 x_2, x_1 x_3$ and $x_2 x_3$.*

The resulting Scarf complex consists of three points (see Fig. 3.4). Again, this is not an acyclic complex.

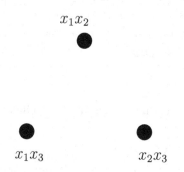

Fig. 3.4. Scarf complex for Example 3.3.11

There is the following obvious, but noteworthy property of the Scarf resolution:

Proposition 3.3.12 *Let $M \subset T = k[x_1^{\pm 1}, \ldots, x_d^{\pm 1}]$ be a monomial ideal. If the Scarf resolution of M exists (that is, if the Scarf complex of M is acyclic), then it is minimal.*

Proof. The condition of the second part of Proposition 3.1.2 is obviously satisfied, which proves minimality.

In [6], Bayer and Sturmfels prove existence and therefore minimality of the Scarf resolution for the class of *generic* co-Artinian monomial modules. Here is the definition:

Definition 3.3.13 *Let $M \subseteq T = k[x_1^{\pm 1}, \dots, x_d^{\pm 1}]$ be a co-Artinian monomial module. Let*

$$\partial_0 : \bigoplus_{m \in \mathrm{MinGen}(M)} Se_m \longrightarrow M$$

be a presentation of M, that is, $\partial(e_m) = m$ for all $m \in \mathrm{MinGen}(M)$.

1. *A binomial $n_m e_m - n_{m'} e_{m'} \in \bigoplus_{m \in \mathrm{MinGen}(M)} Se_m$ is called generic if no variable x_i, $i = 1, \dots, d$ appears in m and m' with the same exponent.*
2. *M is called generic if there exists a basis of generic binomials for $\ker \partial_0$.*

Theorem 3.3.14 (Theorem 2.9 [6]) *Let $M \subset T = k[x_1^{\pm 1}, \dots, x_d^{\pm 1}]$ be a generic co-Artinian monomial module. Then the Scarf complex supports a cellular minimal free $\mathbb{Z}^d$−graded resolution of M.*

In [5] Bayer, Peeva and Sturmfels prove the same result for a more restrictive version of genericity:

Theorem 3.3.15 (Theorem 3.2 [5]) *Let $I \subset S = k[x_1, \dots, x_d]$ be a monomial ideal such that for any $m, m' \in \mathrm{MinGen}(M)$, $m \neq m'$, no variable appears in m and m' with the same non-zero exponent. Then the Scarf complex supports a cellular minimal free $\mathbb{Z}^d$−graded resolution of I.*

The Hull Resolution:

In [6] Bayer and Sturmfels introduce another cellular resolution called the Hull resolution.

Definition 3.3.16   1. *For a co-Artinian monomial module the Hull resolution is the cellular resolution supported on the polytopal $\mathbb{Z}^d$-graded subcomplex X_{Hull} – the Hull complex – of $\mathbb{R}^d$ whose faces are the bounded faces of the polyhedral complex which is the convex hull of $(t^{\alpha_1}, \dots, t^{\alpha_d}) \in \mathbb{R}^d$ for $\underline{x}^\alpha \in M$ and t sufficiently large. (See [6, Theorem 2.3] for details.)*
2. *For a face c of X_{Hull} its degree $\mathrm{lcm}(c)$ is given by α such that $\underline{x}^\alpha$ is the lcm of the monomials in M corresponding to the vertices of the face c.*

Since by [6, Example 3.11] the Hull complex is in general **not** a locally finite polytopal complex, its topology differs from the one induced by the topology of $\mathbb{R}^d$, see [11, Theorem 8.2]. But it is locally finite in the important case of co-Artinian monomial modules defined by lattices [6, Theorem 3.14].

The Hull resolution can be viewed as a generalization of the Scarf resolution, since it always exists and there is the following Theorem:

Theorem 3.3.17 *(See [6], Theorem 2.9) If $M \subset T = k[x_1^{\pm 1}, \dots, x_d^{\pm 1}]$ is generic, the Hull resolution coincides with the Scarf resolution and is therefore minimal.*

The Lyubeznik Resolution:

There is another way to continue the approach taken by the Scarf resolution. Starting from the Taylor resolution we discussed the two strategies of either deleting all simplices with property (3.3.1) or even deleting all simplices with property (3.3.2). What we found in Examples 3.3.9 and 3.3.11 suggests to do "something in between": In Example 3.3.9, deleting all simplices σ with property (3.3.1) was not enough. In Example 3.3.11 deleting even all simplices with property (3.3.2) was too much.

Lyubeznik presents in [33] a nice procedure to delete *some* simplices with property (3.3.2) from the Taylor resolution according to a given linear order on MinGen(M) which he shows always results in an acyclic subcomplex of the Taylor complex and therefore gives rise to a cellular free resolution of M:

Definition 3.3.18 (Lyubeznik resolution) *1. Let $\preceq$ be a linear ordering of* MinGen(M). *Then the Luybeznik complex . is defined to be the simplicial complex on the ground set* MinGen(M) *that consists of all simplices* $\sigma = \{m_0, \ldots, m_s\}$, $m_0 \prec m_1 \prec \ldots \prec m_s$ *that satisfy:*

for all $t < s$ and $m \in$ MinGen(M) s. t. $m \prec m_t$ we have

$$m \nmid \mathrm{lcm}\{m_t, \ldots, m_s\}.$$

2. Lyubeznik's resolution is the $\mathbb{Z}^d$-graded cellular free resolution supported by the Lyubeznik complex.

The above definition says that one gets the Lyubeznik complex from the Taylor complex by deleting those simplices $\sigma = \{m_0, \ldots, m_s\}$, $m_0 \prec m_1 \prec \ldots \prec m_s$ with property (3.3.2) that contain a "tail" $\{m_t, \ldots, m_s\}$ for which there exists a monomial $m \in$ MinGen(M) such that $m \prec m_t$ and $m \mid \mathrm{lcm}\{m_t, \ldots, m_s\}$. Fig. 3.5 shows a Lyubeznik complex for the ideal from Example 3.3.11.

Resolutions Defined by Rooted Complexes:

Recall (see e.g. Example 2.4.2:)

Definition 3.3.19 *A partially ordered P set is called a lattice if for all $x, y \in P$ there exists both*

- *an element $z \in P$ such that $x, y \preceq z$ and for all $u \in P$ such that $x, y \preceq u$ we have $z \preceq u$, called the join $x \vee y$ of x and y, and*
- *an element $z' \in P$ such that $z' \preceq x, y$ and for all $u \in P$ such that $u \preceq x, y$ we have $u \preceq z'$, called the meet $x \wedge y$ of x and y.*

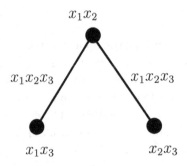

Fig. 3.5. Lyubeznik complex for Example 3.3.11 when using the ordering $x_1x_2 \prec x_1x_3 \prec x_2x_3$

For co-Artinian monomial modules we will frequently make use of the following lattice associated to it:

Definition 3.3.20 *Let $M \subset k[x_1^{\pm 1}, \ldots, x_d^{\pm 1}]$ be a co-Artinian monomial module. We define* LCM(M) *to be the lattice consisting of all* lcm*'s of finite subsets of* MinGen(M).

Note that $1 = \text{lcm}(\emptyset) \in \text{LCM(M)}$ guarantees that this indeed is a lattice.

Novik presents in [41] a generalization of Lyubeznik's resolution which is constructed via *rooted complexes*. These were introduced by Björner and Ziegler in [10].

Definition 3.3.21 *1. A rooting map for a monomial ideal I is a function*

$$\pi : \text{LCM(I)} \longrightarrow \text{MinGen(I)}$$

such that for all $m \in \text{LCM(I)}$ we have

$$\pi(m) \mid m \tag{3.3.3}$$

$$\pi(m) \mid m' \mid m \Rightarrow \pi(m') = \pi(m) \text{ for all } m' \in \text{LCM(I)} \tag{3.3.4}$$

2. Given a rooting map π and a nonempty subset S of MinGen(I), we define

$$\tilde{\pi}(S) := \pi(\text{lcm}(S))$$

We say that a subset S of MinGen(I) *is unbroken if $\tilde{\pi}(S) \in S$. We call S rooted if all nonempty subsets of S are unbroken. The rooted subsets of* MinGen(I) *obviously form a simplicial complex which is called the rooted complex $\mathcal{RC}(I, \pi)$ of I (with respect to the rooting map π).*

Novik proves in [41]:

Theorem 3.3.22 *Let I be a monomial ideal, π a rooting map for I. Then the rooted complex $\mathcal{RC}(I, \pi)$ supports a $\mathbb{Z}^d$-graded cellular free resolution of I.*

This indeed is a generalization of Lyubeznik's construction: Let $\preceq$ be a linear order on $\mathrm{MinGen}(I)$. Define a rooting map by setting

$$\pi(m) := \min_{\preceq}\left\{m' \in \mathrm{MinGen}(I) \mid m' \big| m\right\}. \tag{3.3.5}$$

It is easy to see that under this choice of the rooting map π the rooted complex $\mathcal{RC}(I, \pi)$ coincides with the Lyubeznik complex.

Novik also presents a condition under which the resolution given by rooted complexes is minimal.

For the statement we need the following definition:

Definition 3.3.23 *A finite lattice L is called geometric if it is atomic , that is, every element is a join of atoms, and semimodular, which means that if for elements $x, y \in L$ there exists no element $z \in L$ such that $x \wedge y < z < x$ then there also exists no element $z \in L$ such that $y < z < x \vee y$.*

Theorem 3.3.24 *Let I be a monomial ideal, such that $\mathrm{LCM}(I)$ is a geometric lattice and let π be a rooting map for I. Then the rooted complex $\mathcal{RC}(I, \pi)$ supports a cellular minimal free resolution of I.*

The LCM- and LCM^*-Resolutions:

In this section we introduce cellular resolutions of monomial modules derived from their LCM-lattices. This is a special case of resolutions obtained from partial orders on monomials that appeared first in [46].

Consider the following example:

Example 3.3.25 *Let $M \subset k[x_1, \ldots, x_d]$ be the monomial module generated by all d monomials of the form $x_1 \ldots x_{i-1}x_{i+1} \ldots x_d$, $i = 1, \ldots, d$. There is a cellular resolution of M given by a one-dimensional simplicial complex consisting of $d + 1$ cells of dimension 0 and d cells of dimension 1.*

Proof. Consider the following $\mathbb{Z}^d$-graded simplicial complex X: There are $d+1$ points, the first d of them correspond to the minimal generators of M and are graded accordingly. The last point is graded by $(1, \ldots, 1) \in \mathbb{Z}^d$. For each of the first d points there is a line connecting it with the last point. All these lines are graded by $(1, \ldots, 1) \in \mathbb{Z}^d$. It is easily seen that this graded simplicial complex satisfies the condition of Proposition 3.1.2 since for all $\alpha \in \mathbb{Z}^d$ the subcomplex $X_{\leq \alpha}$ is empty, consists of one point or equals the whole complex X.

Note that the above resolution has one 0-cell in addition to the ones that come from the Taylor resolution. Therefore, the resolution does not qualify for minimality, even though it has much fewer cells of all other dimensions. The above resolution is an example of the following class of resolutions which we will call LCM-resolutions:

Definition 3.3.26   *1. Let P be a partially ordered set. The simplicial complex $\Delta(P)$ that consists of all finite chains in P, that is all finite subsets of P that are totally ordered by the partial order of P, is called the order complex of P.*

2. Let $M \subset k[x_1^{\pm 1}, \ldots, x_d^{\pm 1}]$ be a co-Artinian monomial module. Denote by $X_{\mathrm{LCM}}(M)$ the order complex of $\mathrm{LCM}(M) - \{1\}$ (see Definition 3.3.20 for the definition of $\mathrm{LCM}(M)$). We refer to this complex as the LCM-complex of M. We grade the simplices of $X_{\mathrm{LCM}}(M)$ in the obvious way by lcm.

Proposition 3.3.27 *Let M be a co-Artinian monomial module. Then there is a $\mathbb{Z}^d$-graded cellular free resolution of M supported by $X_{\mathrm{LCM}}(M)$.*

We call this resolution the LCM-*resolution of M.*

Proof. For all multidegrees $\alpha \in \mathbb{Z}^d$, that appear as an lcm, there exists a unique 0-chain $\{\mathbf{x}^\alpha\}$ graded with α. For all simplices σ of the subcomplex $(X_{\mathrm{LCM}}(M))_{\leq \alpha}$ one of the following holds:

1. $\mathbf{x}^\alpha \in \sigma$ and $\sigma - \{\mathbf{x}^\alpha\} \in (X_{\mathrm{LCM}}(M))_{\leq \alpha}$,
2. $\mathbf{x}^\alpha \notin \sigma$ and $\sigma \cup \{\mathbf{x}^\alpha\} \in (X_{\mathrm{LCM}}(M))_{\leq \alpha}$.

Hence $(X_{\mathrm{LCM}}(M))_{\leq \alpha}$ forms a cone, therefore is contractible and acyclic. The assertion follows with Proposition 3.1.2.

We now introduce another class of cellular resolutions which is similar to the class of LCM-resolutions:

Definition 3.3.28 *Let $M \subset k[x_1^{\pm 1}, \ldots, x_d^{\pm 1}]$ be a co-Artinian monomial module. Let $\mathrm{LCM}^*(M)$ be the partially ordered set consisting of all monomials in M that divide lcm σ for some finite subset $\sigma \subset \mathrm{MinGen}(M)$. We set $\Delta(\mathrm{LCM}^*(M))$ to be the order complex of $\mathrm{LCM}^*(M)$ graded by lcm in the obvious way. We refer to this complex as the LCM^*- complex.*

Note that $\mathrm{LCM}^*(M)$ is the convex hull of $\mathrm{LCM}(M) - \{1\}$ in the set of all monomials in $k[x_1^{\pm 1}, \ldots, x_d^{\pm 1}]$ in the sense that we get $\mathrm{LCM}^*(M)$ from $\mathrm{LCM}(M) - \{1\}$ by adding all monomials $m \in k[x_1^{\pm 1}, \ldots, x_d^{\pm 1}]$ with the property that there exist monomials $m_1, m_2 \in \mathrm{LCM}(M)$ such that $m_1 \big| m \big| m_2$.

Proposition 3.3.29 *Let M be a co-Artinian monomial module. Then there is a $\mathbb{Z}^d$-graded cellular free resolution of M supported by $\Delta(\mathrm{LCM}^*(M))$.*

Proof. The proof is the same as for Proposition 3.3.27.

The Hypersimplex Resolution:

In this subsection we present a $\mathbb{Z}^d$-graded cellular free resolution of the powers $\mathfrak{m}^n$ of the maximal homogeneous ideal $\mathfrak{m} = (x_1, \ldots, x_d) \subset S = k[x_1, \ldots, x_d]$. This approach was first developed and presented in [3].

Definition 3.3.30 *Let $\mathcal{C}_d^n$ be the polytopal CW-complex with*

$$\Delta_n := n \cdot \Delta^{d-1} = \left\{ (y_1, \ldots, y_d) \in \mathbb{R}^d \mid \sum_{i=1}^{d} y_i = n, \ y_i \geq 0, \ i = 1, \ldots, d \right\}$$

as underlying space and CW-complex-structure induced by intersection with the cubical CW-complex-structure on $\mathbb{R}^d$ given by the integer lattice $\mathbb{Z}^d$. That is, the closed cells of $\mathcal{C}_d^n$ are given by all hypersimplices

$$C_{\underline{a},J} := \Delta_n \cap \left\{ (y_1, \ldots, y_d) \in \mathbb{R}^d \mid y_i = a_i \ : i \in [d] \backslash J, \ y_j \in [a_j, a_j+1] \ : j \in J \right\}$$

$$= \mathrm{Conv}\left(\underline{a} + \sum_{j \in J} \epsilon_j e_j \mid \epsilon_j \in \{0,1\}, \ \sum_{j \in J} \epsilon_j = d - |\underline{a}| \right)$$

with $\underline{a} \in \mathbb{N}^d$, $J \subset [d] := \{1, \ldots, d\}$, $|\underline{a}| := \sum_{i \in [d]} a_i$, e_i the i-th unit vector in $\mathbb{R}^d$, either subject to the conditions $|\underline{a}| = n$ and $J = \emptyset$, (these are the $0-$cells $C_{\underline{a},\emptyset} = \{\underline{a}\}$,) or the condition $1 \leq n - |\underline{a}| \leq |J| - 1$. The CW-complex $\mathcal{C}_d^n$ is naturally multigraded by setting $\mathrm{lcm}(C_{\underline{a},J}) := \underline{a} + \sum_{j \in J} e_j$. We refer to $\mathcal{C}_d^n$ as the hypersimplicial complex.

An example of a hypersimplicial complex is shown in Fig. 3.6.

Lemma 3.3.31 *Considering canonical orientations of these hypersimplices and denoting $J = \{j_0, \ldots, j_r\}$, $j_0 < \ldots < j_r$, $J_\nu := J \backslash \{j_\nu\}$, the differential of $\mathcal{C}_d^n$ is given by*

$$\bullet \ \partial C_{\underline{a},J} = \sum_{\nu=0}^{r} (-1)^\nu (C_{\underline{a},J_\nu} - C_{\underline{a}+e_{j_\nu},J_\nu}), \quad \text{if } 2 \leq n - |\underline{a}| \leq |J| - 2,$$

$$\bullet \qquad \partial C_{\underline{a},J} = \sum_{\nu=0}^{r} (-1)^\nu C_{\underline{a},J_\nu}, \qquad \text{if } 1 = n - |\underline{a}| \leq |J| - 2,$$

$$\bullet \qquad \partial C_{\underline{a},J} = \sum_{\nu=0}^{r} (-1)^{\nu+1} C_{\underline{a}+e_{j_\nu},J_\nu}, \qquad \text{if } 2 \leq n - |\underline{a}| = |J| - 1,$$

$$\bullet \qquad \partial C_{\underline{a},\{j_0,j_1\}} = C_{\underline{a}+e_{j_1},\emptyset} - C_{\underline{a}+e_{j_0},\emptyset}, \qquad \text{if } n - |\underline{a}| = 1,$$

$$\bullet \qquad C_{\underline{a},\emptyset} = 0, \qquad \text{if } |\underline{a}| = n.$$

Proposition 3.3.32 C_d^n *defines a multigraded cellular free resolution of* $\mathfrak{m}^n$.

Proof. The 0-cells of C_d^n are in one-to-one-correspondence with the set of minimal generators of $\mathfrak{m}^n$, so the fact that $(C_d^n)_{\leq\alpha} = \Delta_n \cap \{ (y_1, \ldots, y_d) \in \mathbb{R}^d \mid y_i \leq \alpha_i, i = 1, \ldots, d \}$ is contractible or empty for all $\alpha \in \mathbb{Z}^d$ proves (use Proposition 3.1.2) the assertion.

We would like to mention that recently using a different approach Sinefakopoulos [52] has constructed a cellular minimal free resolution of powers of the maximal ideal that is supported on a shellable regular CW-complex.

The Bar Resolution:

Definition 3.3.33 *Let* $k[\Lambda]$ *be an affine semigroup ring. We set* $\Lambda_0 := \Lambda \setminus \{\underline{0}\}$. *The (normalized) Bar resolution of* k *as an* R- *module is the multigraded resolution*

$$\mathcal{F} = \cdots \xrightarrow{\partial_{i+1}} \mathcal{F}_i \xrightarrow{\partial_i} \mathcal{F}_{i-1} \xrightarrow{\partial_{i-1}} \cdots \xrightarrow{\partial_1} \mathcal{F}_0,$$

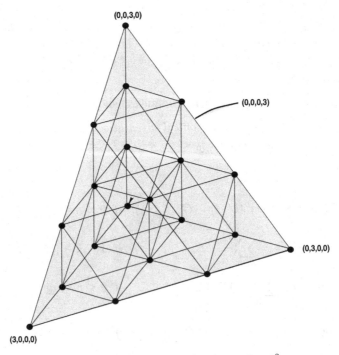

Fig. 3.6. The hypersimplicial complex C_4^3

$\mathcal{F}_i = \bigoplus_{\alpha \in \mathbb{Z}^d} F_i^\alpha$ *such that* F_i^α *is the free R-module with basis given by the set*

$$\Lambda_i^\alpha := \{[\lambda_0|\cdots|\lambda_i] \mid \lambda_j \in \Lambda_0,\ 0 \le j \le i,\ \lambda_0 + \cdots + \lambda_i = \alpha\}.$$

The differential is given by

$$\partial_i[\lambda_0|\cdots|\lambda_i] = \underline{\mathbf{x}}^{\lambda_0}[\lambda_1|\cdots|\lambda_i] + \sum_{j=1}^{i}(-1)^j[\lambda_0|\cdots|\lambda_{j-1} + \lambda_j|\cdots|\lambda_i].$$

In order to see that this construction defines a $\mathbb{Z}^d$-graded free resolution of k, consider the following Lemma:

Lemma 3.3.34 *The Bar resolution is cellular. It is supported by the* $(\mathbb{Z}^d, \Lambda)$-*graded order complex* $\Delta(\Lambda_0)$ *of* Λ_0.

Proof. For $\sigma = (\lambda_0, \dots, \lambda_i) \in \Delta(\Lambda)^{(*)}$, that is $\lambda_0 \prec \dots \prec \lambda_i$, we set $\mathrm{gr}(\sigma) := \lambda_i$. By Definition 2.1.2 it is obvious that this indeed provides a $(\mathbb{Z}^d, \Lambda)$-grading for the order complex $\Delta(\Lambda_0)$ of Λ_0. We set $\Delta(\Lambda_0)_\alpha^{(i)} := \{\sigma \in \Delta(\Lambda_0)^{(i)} \mid \mathrm{gr}(\sigma) = \alpha\}$. Consider the following maps

$$\Phi_\alpha^i : \quad \Lambda_i^\alpha \quad \longrightarrow \Delta(\Lambda_0)_\alpha^{(i)}$$
$$[\lambda_0, \dots, \lambda_i] \longmapsto (\lambda_i, \lambda_{i-1} + \lambda_i, \dots, \lambda_0 + \dots + \lambda_i).$$

For all $i \in \mathbb{N}$ the map Φ_α^i provides a one-to-one correspondences between Λ_i^α and $\Delta(\Lambda_0)_\alpha^{(i)}$. The differential of the simplicial complex $\Delta(\Lambda_0)$ induces the following map:

$$[\lambda_0|\cdots|\lambda_i] \longmapsto [\lambda_1|\cdots|\lambda_i] + \sum_{j=1}^{i}(-1)^j[\lambda_0|\cdots|\lambda_{j-1} + \lambda_j|\cdots|\lambda_i].$$

The given differential comes from homogenizing this map. This proves the assertion.

Exercises

Exercise 3.3.1. Given an arrangement of hyperplanes $\mathcal{A}$ corresponding the matroid M. Use the CW-complex from Example 2.2.7 to give a minimal free resolution of $k[M]$. (This is a very hard exercise. The construction first appeared in [42].)

Exercise 3.3.2. Show that if I is a monomial ideal such that $\gcd(m, n) = 1$ for any two monomials m and n in the minimal monomial generating set of I then the Taylor resolution of I is minimal.

Exercise 3.3.3. The term generic co-Artinian monomial module suggests that a 'randomly' chosen co-Artinian monomial module is generic with probability 1. Justify the terminology 'generic' for the class of monomial ideals satisfying the conditions of Theorem 3.3.15.

Exercise 3.3.4. Give conditions on a monomial ideal I such that the Taylor and the Scarf resolution coincide.

Exercise 3.3.5. Make the term 'sufficiently large' in Definition 3.3.16 precise. (see also [6]).

Exercise 3.3.6. Describe the CW-complex underlying the hull resolution of the Stanley-Reisner ideal of the uniform matroid $U_{n,l}$.

Exercise 3.3.7. Give conditions on a monomial ideal such that for a suitable choice of $\preceq$ in Definition 3.3.18 the Lyubeznik resolution is minimal (see also [33]).

Exercise 3.3.8. Find a monomial ideal for which LCM(I) is not geometric but the rooted complex $\mathcal{RC}(I, \pi)$ supports a cellular minimal free resolution of I.

Exercise 3.3.9. Describe the LCM-lattices of the powers $\mathfrak{m}^\ell$, $\ell \geq 1$, of the maximal homogeneous ideal in $k[x_1, \ldots, x_d]$.

Exercise 3.3.10. Classify the monomial ideals for which the LCM-lattice is isomorphic to the lattice of subsets of an n-element set.

4

Discrete Morse Theory

This chapter contains a presentation of discrete Morse theory as developed by Robin Forman (see e.g. [20], [21]). This theory allows to combinatorially construct from a given (regular, finite) CW-complex a second CW-complex that is homotopy equivalent to the first but has fewer cells. As the upshot of this chapter we then show that one can use this theory in order to construct minimal free resolutions (see also [3]). Discrete Morse theory has found many more applications in Geometric Combinatorics and other fields of mathematics, we will not be able to speak about them. We refer the reader for example to [29] where most applications of discrete Morse theory to complexes of graphs are reviewed. There are even promising attempts to find real world applications of discrete Morse theory (see [31]) to image analysis.

The applications to free resolutions require (e.g. for the Bar resolution) an extension of Forman's theory to not necessarily finite graded (regular) CW-complexes in Section 4.2. Then in Section 4.3 we present and develop the application of discrete Morse theory to cellular resolutions. Since for the construction of minimal free resolutions it is very often important and interesting to control the differential of the chain complex, we also must control the differential of the CW-complexes constructed by discrete Morse theory. Therefore, Section 4.4 discusses Morse differentials.

Except for some exercises and questions we do not give actual applications of discrete Morse theory to the construction of minimal free resolutions. Such applications can for example be found in [3], [4] and [24]. Indeed, even though it has turned out that the method is quite powerful, most applications are rather technical. Therefore, we think that they are not well suited for these lecture notes.

4.1 Forman's Discrete Morse Theory

The fundamental tool from Combinatorial Topology applied in these notes is *discrete Morse theory* as developed by Forman [20, 21]. In this section, we review parts of this theory.

Definition 4.1.1 *Let X be a CW-complex, $\sigma, \tau \in X^{(*)}$ cells in X such that σ is a face of τ, $\dim(\sigma) = i - 1$, $\dim(\tau) = i$. Let*

$$f_\sigma : B^{i-1} \to X \text{ and } f_\tau : B^i \to X$$

be the characteristic maps for σ and τ. We say that σ is a regular face of τ if there exist homeomorphisms

$$g_\tau : B^i \xrightarrow{\cong} B^i$$

and

$$g_\sigma : S^{i-1}_{\geq 0} := \{(x_1, \ldots, x_i) \in S^{i-1} \mid x_1 \geq 0\} \xrightarrow{\cong} B^{i-1}$$

such that the following diagram commutes:

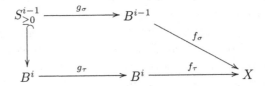

The central idea of discrete Morse theory is to construct for a given CW-complex a new CW-complex that is homotopically equivalent to the original one but is built from fewer cells. The condition needed for such a construction is the existence of a *discrete Morse function* .

Definition 4.1.2 (Discrete Morse Function) *A discrete Morse function on the CW-complex X is a function*

$$f : X^{(*)} \longrightarrow \mathbb{R}$$

such that

1. *for every cell $\sigma \in X^{(*)}$*

$$| \{\tau \in X^{(*)} \mid \tau \text{ is a facet of } \sigma, \ f(\tau) \geq f(\sigma)\} | \leq 1 \qquad (4.1.1)$$
$$| \{\tau \in X^{(*)} \mid \sigma \text{ is a facet of } \tau, \ f(\tau) \leq f(\sigma)\} | \leq 1 \qquad (4.1.2)$$

and

2. if $\sigma, \tau \in X^{(*)}$ are cells in X, σ a facet of τ , such that

$$f(\sigma) \geq f(\tau),$$

then σ is a regular facet of τ.

Example 4.1.1. All functions on $X^{(*)}$ that increase strictly with dimension are discrete Morse functions.

Discrete Morse functions can be regarded as functions that increase with dimension up to one exception locally.

Definition 4.1.3 (Critical Cells) *Let X be a finite CW-complex, f a discrete Morse function on X. A cell σ of X is called f-critical , if*

$$| \{\tau \in X^{(*)} \mid \tau \text{ is a facet of } \sigma, \ f(\tau) \geq f(\sigma)\} | = 0 \qquad (4.1.3)$$
$$| \{\tau \in X^{(*)} \mid \sigma \text{ is a facet of } \tau, \ f(\tau) \leq f(\sigma)\} | = 0. \qquad (4.1.4)$$

We set

$$X^{(*)}_{critical}(f) := \{\sigma \in X^{(*)} \mid \sigma \text{ is } f\text{-critical}\}$$

For every dimension i the number of f-critical cells in dimension i is called the Morse number $m_i(f)$.

The key result by Forman (see [20], Theorem 10.2), that lays the ground for all subsequent developments, is given in the following theorem. Note that Forman's formulation requires a regular CW-complex. But indeed he only needs some local regularity assumptions that are build in our discrete Morse functions (see last part of Definition 4.1.2).

Theorem 4.1.4 *Let X be a finite CW-complex, f a discrete Morse function on X. Then X is homotopy equivalent to a CW-complex with exactly $m_i(f)$ cells of dimension i.*

Example 4.1.2. If f is a discrete Morse function from Example 4.1.1 then all cells are critical and hence Theorem 4.1.4 becomes trivial. Based on this observation we will call these discrete Morse functions *trivial* .

Exercises

Exercise 4.1.3. Show using discrete Morse theory that the full simplex 2^{Ω} over some finite ground set Ω is contractible.

Exercise 4.1.4. A face A of a simplicial complex Δ is called free , if A is not maximal and contained in a unique maximal face. Show using discrete Morse theory that Δ and $\Delta \setminus \{B \mid A \subseteq B \in \Delta\}$ are homotopy equivalent.

Exercise 4.1.5. Show that given a function $f : X^{(0)} \to \mathbb{R}$ from the 0-cells of a regular CW-complex X to the reals it is always possible to extend f to a discrete Morse function on X. (See [31] for clever ways to do so.)

4.2 Discrete Morse Theory for Graded CW-Complexes

In this section we expand discrete Morse theory to graded and not necessarily finite CW-complexes. In [14], Chari reformulates discrete Morse theory in terms of acyclic matchings. In this section we review his results. On the way we show that his approach leads in a natural way to the consideration of gradings by partially ordered sets. As the main result of this section (see Theorem 4.2.14) we prove a version of Theorem 4.1.4 for graded CW-complexes.

We begin with the basic definitions from Chari's [14] approach to discrete Morse theory.

Definition 4.2.1 (Cell Graph) *Let X be a CW-complex. Consider the directed graph G_X on $X^{(*)}$ whose set E_X of edges is given by*

$$E_X := \{\tau \to \sigma \mid \sigma \text{ is a facet of } \tau\}.$$

We call $G_X = (X^{(*)}, E_X)$ the cell graph of X.

Definition 4.2.2 (Acyclic Matching) *Let X be a CW-complex, $G_X = (X^{(*)}, E_X)$ its cell graph. Let $A \subset E_X$ be a subset of edges $\tau \to \sigma \in E_X$, such that σ is a regular face of τ for all $\tau \to \sigma \in E_X$.*

1. We denote by $G_X^A = (X^{(*)}, E_X^A)$ the induced graph with edge set

$$E_X^A := (E_X \setminus A) \cup \{\sigma \to \tau \mid \tau \to \sigma \in A\}$$

that is built from G_X by reversing the direction of all edges $\tau \to \sigma$ that lie in A. We call an edge $\sigma \to \tau \in E_X^A$ an A-edge if for its reversed edge $\tau \to \sigma$ we have

$$\tau \to \sigma \in A.$$

2. We call A a matching on X, if each cell $\sigma \in X^{(*)}$ occurs in at most one edge of A.

3. We call A an acyclic matching on X, if A is a matching and if the induced graph G_X^A is acyclic, that is, contains no directed cycle.

4. A cell of X is called A-critical if it does not occur in any edge $\tau \to \sigma \in A$.
5. We set

$$X_{critical}^{(*)}(A) := \{\sigma \in X^{(*)} \mid \sigma \text{ is } A\text{-critical}\}$$

6. We denote by $\tilde{G}_X^A = (X^{(*)}, \tilde{E}_X^A)$ the induced graph with edge set

$$\tilde{E}_X^A := E_X \cup \{\sigma \to \tau \mid \tau \to \sigma \in A\}$$

that is built from G_X by adding for all edges $\tau \to \sigma$ that lie in A their reversed edges $\sigma \to \tau$.

Example 4.2.1. Let $\Delta = 2^{[3]}$ be the 2-simplex on ground set $[3]$. The following figure shows an acyclic and a non-acyclic matching on G_Δ.

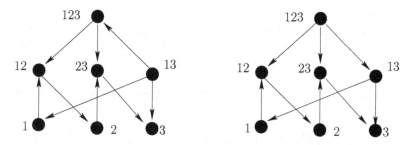

Fig. 4.1. Cell graphs for Example 4.2.1

The following simple lemma shows that one can regard the set of acyclic matchings as a simplicial complex itself (see [15] for results exploiting this fact).

Lemma 4.2.3 *Let X be a CW-complex, A an acyclic matching on X and $A' \subset A$. Then A' is an acyclic matching on X.*

Proof. Suppose

$$\gamma = v_0 \to \ldots \to v_k, \ v_k = v_0$$

is a cycle in A'. Note that, since A' is a matching, no two consecutive edges $v_i \to v_{i+1}$ and $v_{i+1} \to v_{i+2}$ can be A'-edges. From the fact that the dimension of the cells involved decreases strictly along edges that are not A'-edges and increases by 1 along A'-edges it follows that

$$\gamma = \tau_0 \to \sigma_0 \to \ldots \to \tau_k \to \sigma_k \to \tau_{k+1}, \ \tau_{k+1} = \tau_0,$$

where all

$$\sigma_i \to \tau_{i+1}, \ i = 0, \ldots, k$$

are A'-edges and all

$$\tau_i \to \sigma_i, \ i = 0, \ldots, k$$

are not A'-edges (or the other way around). Hence γ is also a cycle in G_X^A, since none of the edges involved that are not A'-edges can be an A-edge (this would violate the property of A of being a matching). This completes the proof and shows that A' must be acyclic.

Lemma 4.2.4 *Let X be a CW-complex, A an acyclic matching on X. Then the only cycles in the graph $\tilde{G}_X^A$ consist of exactly one edge $e \in A$ and its reversed edge.*

Proof. If γ is a cycle in $\tilde{G}_X^A$, it must contain one edge $e \in A$ and its reversed edge. Assume not. Then γ can be viewed as a cycle in $G_X^{A'}$ for some $A' \subset A$. But this conflicts with Lemma 4.2.3.

These two edges cut γ into two parts that again are cycles, so by induction hypothesis γ only consists of pairs of reversely directed edges. From the fact that A is a matching it follows that γ consists of only one such pair. $\square$

The preceding lemma leads us to the definition of a poset structure on $A^{(*)}$.

Definition 4.2.5 (The Matching Poset $(A^{(*)}, \preceq_A)$) *Let X be a CW-complex, A an acyclic matching on X.*

1. *We define*

$$A^{(*)} := A \cup X_{critical}^{(*)}(A).$$

   For $a, b \in A^{(*)}$, we set $a \preceq_A b :\Leftrightarrow$ there exists a path in $\tilde{G}_X^A$ from b to a. Here, if $a = \tau \to \sigma \in A$, we mean a path in $\tilde{G}_X^A$ from b to either σ or τ, if $b = \tau \to \sigma \in A$, we mean a path in G_X^A from either σ or τ to a. We call the partially ordered set $(A^{(*)}, \preceq_A)$ the matching poset of A.

2. *We call the function*

$$\mathrm{gr}_A : X^{(*)} \longrightarrow A^{(*)}$$

$$\sigma \longmapsto \begin{cases} \sigma & \text{if } \sigma \in X_{critical}^{(*)} \\ \tau' \to \tau & \text{if } \sigma \in \{\tau, \tau'\} \text{ and } \tau' \to \tau \in A \end{cases}$$

   the A-universal grading of X. We denote the subcomplexes $X_{\preceq a}$ that arise from this $A^{(*)}$-grading by $X_{\preceq_A a}$.

Example 4.2.2. Let Δ_n be the simplicial complex on ground set $\Omega_n = \{\pm 1, \ldots, \pm n\}$ where $B \subseteq \Omega_n$ lies in Δ_n if and only if $i \in B$ implies $-i \notin B$ – as usual $--i = i$. Note that the boundary complex of the n-dimensional hyper-octahedron is a geometric realization of Δ_n. Consider the matching A

on G_{Δ_n} which contains all edges $B \to B \setminus \{n\}$ for $n \in B \in \Delta_n$, $B \neq \{n\}$..
Then A is an acyclic matching. Its matching poset consists of all faces of Δ_n
which contain $-n$ and the face $\{n\}$ together with the edges from A. For the
order relation see Exercise 4.2.3.

The following definitions and lemmas clarify the relation between discrete
Morse functions and acyclic matchings.

Definition 4.2.6 *Let X be a CW-complex, $G_X = (X^{(*)}, E_X)$ its cell graph
and $f : X^{(*)} \longrightarrow \mathbb{R}$ a discrete Morse function on X. We set*

$$A_f := \{\tau \to \sigma \in E_X \mid \sigma \text{ is a facet of } \tau, \ f(\tau) \leq f(\sigma)\}.$$

We call A_f the acyclic matching on X corresponding to f.

The next lemmas explain the preceding definition and link Forman's original approach to the one taken by Chari.

Lemma 4.2.7 *Let X be a CW-complex, $f : X^{(*)} \longrightarrow \mathbb{R}$ a discrete Morse
function on X. Then*

1. *A_f is an acyclic matching and*
2. *$X_{critical}^{(*)}(A_f) = X_{critical}^{(*)}(f)$.*

Lemma 4.2.7 and the subsequent Lemma 4.2.8 also justify our choice of
notation $X_{\text{critical}}^{(*)}$ for the set of critical cells in a context in which a single acyclic
matching and its corresponding discrete Morse functions are involved.

Proof. 1. By [20, Lemma 2.5] it is impossible to violate conditions (4.1.3)
   and (4.1.4) of Definition 4.1.3 simultaneously for a single cell $\sigma \in X^{(*)}$.
   Therefore, all cells $\sigma \in X^{(*)}$ occur in at most one edge of A_f. This in turn
   shows that A_f is a matching. Acyclicity of A_f is derived from the fact
   that f decreases on paths in G_X^A and even decreases strictly along edges
   $\sigma \to \tau$ that are not A-edges. From the latter edges at least one must occur
   in any cycle.
2. This is obvious from the construction of A_f.

Lemma 4.2.8 *Let X be a CW-complex, A an acyclic matching on X. Then
there exists a discrete Morse function on X such that*

$$A = A_f.$$

Proof. Let $\prec$ be any linear extension of $\prec_A$ and

$$f^{(*)} : A^{(*)} \longrightarrow \mathbb{R}$$

strictly $\prec$-order preserving, that is, for $a, b \in A^{(*)}$ we have

$$a \prec b \Longrightarrow f^{(*)}(a) < f^{(*)}(b).$$

We define a discrete Morse function $f : X^{(*)} \longrightarrow \mathbb{R}$ by

$$f(\sigma) = \begin{cases} f^{(*)}(a) & \text{if } \sigma = a \in X^{(*)}_{\text{critical}} \\ f^{(*)}(a) & \text{if } a = \tau \to \upsilon \in A \text{ and } \sigma \in \{\tau, \upsilon.\} \end{cases}$$

It is easy to check that f is a discrete Morse function on X and $A_f = A$.

We summarize the link between Forman's and Chari's approach provided by Lemma 4.2.7 and Lemma 4.2.8 in the following reformulation of Theorem 4.1.4. Again note that in contrast to Forman and Chari, we do not require regularity of the full CW-complex but have build in local regularity assumptions in Definitions 4.1.2 and 4.2.2.

Theorem 4.2.9 *Let X be a finite CW-complex, A an acyclic matching on X. Then there is a CW-complex X_A whose i-cells are in one-to-one correspondence with the A-critical i-cells of X such that X_A is homotopy equivalent to X.*

In the remainder of this section we give an explicit description of the Morse complex X_A corresponding to an acyclic matching A on X, extend the theory to not necessarily finite CW-complexes and show that the Morse equivalence is, in a canonical way, compatible with gradings.

Definition 4.2.10 *Let P be a partially ordered set, (X, gr) a P-graded CW-complex.*

1. *We call the grading $\mathrm{gr} : X^{(*)} \longrightarrow P$ a compact P-grading of X if $X_{\preceq p}$ is compact for all $p \in P$. X is then called compactly P-graded.*
2. *We call an acyclic matching A of X proper if the corresponding A-universal grading $\mathrm{gr}_A : X^{(*)} \longrightarrow A^{(*)}$ is compact.*

In the sequel we provide prerequisites for the construction of the Morse complex X_A corresponding to a proper acyclic matching A on X.

Lemma 4.2.11 *Let X be a CW-complex.*

1. *Let $\sigma, \tau \in X^{(*)}$, $\dim = i$, $\dim \tau = i+1$, σ a regular face of τ. Then there exists a deformation retraction*

$$h_{\tau \to \sigma} : \bar{\tau} \longrightarrow \bigcup_{\substack{\sigma' \in X^{(*)}, \\ \sigma' \text{ a face of } \tau, \\ \sigma' \notin \{\sigma, \tau\}}} \overline{\sigma'}.$$

2. Let A be an acyclic matching on G_X, $a \in A^{(*)}$. If $a = (\tau \to \sigma) \in A$, then

$$\tilde{h}_{\tau \to \sigma} : X_{\preceq_A a} \longrightarrow X_{\prec_A a}$$

$$x \longmapsto \begin{cases} x & \text{if } x \notin \bar{\tau} \\ h_{\tau \to \sigma}(x) & \text{if } x \in \bar{\tau}, \end{cases}$$

is a deformation retraction.

Proof. The assertions follows immediately from Definition 4.1.1

Definition 4.2.12 *Let A, X, Y and Y' be topological spaces, $A \subset X$, $\phi : Y \longrightarrow Y'$ and $f : A \longrightarrow Y$ continuous maps. We denote by*

$$\text{id} \cup_f \phi : X \cup_f Y \longrightarrow X \cup_{\phi \circ f} Y'$$

the unique map that makes the following diagram commutative:

$$
\begin{array}{ccc}
X \dot\cup Y & \xrightarrow{\ \text{id}\dot\cup\phi\ } & X \dot\cup Y' \\
\downarrow & & \downarrow \\
X \cup_f Y & \xrightarrow{\ \text{id}\cup_f\phi\ } & X \cup_{\phi\circ f} Y'
\end{array}
$$

The following definition provides the construction of the Morse complex.

Definition 4.2.13 (Morse complex, Morse equivalence) *Let X be a CW-complex, A a proper acyclic matching on X.*

Recall (see Definition 4.2.5) that $\preceq_A$ denotes the partial order of the matching poset $A^{(*)}$. For all $a \in A^{(*)}$ we define inductively a CW-complex $(X_A)_{\preceq_A a}$, $a \in A^{(*)}$ and a map

$$H(A)_{\preceq_A a} : X_{\preceq_A a} \longrightarrow (X_A)_{\preceq_A a} :$$

1. Let $a \in A^{(*)}$ be $\prec_A$-minimal. We set

$$(X_A)_{\preceq_A a} := X_{\preceq_A a} = a$$

(note that $a \prec_A$-minimal $\Rightarrow a \in X^{(*)}_{critical}$) and

$$H(A)_{\preceq_A a} := id_{X_{\preceq_A a}}.$$

2. Let $a \in A^{(*)}$. Suppose for all $b \prec_A a$, the CW-complexes $(X_A)_{\preceq_A b}$ and the maps $H(A)_{\preceq_A b} : X_{\preceq_A b} \longrightarrow (X_A)_{\preceq_A b}$ are constructed such that for all $b, b' \in A^{(*)}$, $b' \preceq_A b \prec_A a$, we have

$$(X_A)_{\preceq_A b'} \subset (X_A)_{\preceq_A b}$$

and the diagram

$$
\begin{array}{ccc}
X_{\preceq_A b'} & \xrightarrow{\;H(A)_{\preceq_A b'}\;} & (X_A)_{\preceq_a b'} \\
\Big\uparrow & & \Big\uparrow \\
X_{\preceq_A b} & \xrightarrow{\;H(A)_{\preceq_A b}\;} & (X_A)_{\preceq_A b}
\end{array}
$$

commutes.

Denote by $(X_A)_{\prec_A a}$ *the union*

$$
(X_A)_{\prec_A a} := \bigcup_{b \prec_A a} (X_A)_{\preceq_A b}
$$

and by

$$
H(A)_{\prec_A a} : X_{\prec_A a} \longrightarrow (X_A)_{\prec_A a}
$$

the map induced by the maps

$$
H(A)_{\preceq_A b} : X_{\preceq_A b} \longrightarrow (X_A)_{\preceq_A b}, \; b \prec_A a.
$$

- *Case 1:* $a = \tau \to \sigma \in A$

  We set
  $$
  (X_A)_{\preceq_A a} := (X_A)_{\prec_A a}
  $$

  and define
  $$
  H(A)_{\preceq_A a} : X_{\preceq_A a} \longrightarrow (X_A)_{\preceq_A a}
  $$

  by
  $$
  H(A)_{\preceq_A a} := H(A)_{\prec_A a} \circ \tilde{h}_{\tau \to \sigma}.
  $$

- *Case 2:* $a = \sigma \in X^{(*)}_{critical}$

  Let $\dim \sigma = i$ *and* $f_\sigma : B^i_\sigma \longrightarrow X_{\preceq_A a}$ *be the characteristic map of the cell* σ*. Recall that*
  $$
  X_{\preceq_A a} = B^i_\sigma \cup_{f_{\partial \sigma}} X_{\prec_A a}.
  $$

  We set
  $$
  (X_A)_{\preceq_A a} := B^i_\sigma \cup_{H(A)_{\prec_A a} \circ f_{\partial \sigma}} (X_A)_{\prec_A a}
  $$

  and
  $$
  H(A)_{\preceq_A a} := \mathrm{id}_{B^i_\sigma} \cup_{f_{\partial \sigma}} H(A)_{\prec_A a}.
  $$

We define

$$
X_A := \bigcup_{a \in A^{(*)}} (X_A)_{\preceq_A a}
$$

and

$$H(A) : X \longrightarrow X_A$$

to be the map induced by the maps

$$H(A)_{\preceq_A a} : X_{\preceq_A a} \longrightarrow (X_A)_{\preceq_A a}.$$

We call X_A the Morse complex of X with respect to A and $H(A)$ the corresponding Morse equivalence.

Theorem 4.2.14 *Let X be a CW-complex, A a proper acyclic matching on X. Let X_A be the corresponding Morse complex, $H(A) : X \longrightarrow X_A$ the corresponding Morse equivalence. Then:*

1. *For all $i \in \mathbb{N}$, the i-cells of X_A are in one-to-one correspondence with the A-critical i-cells of X.*
2. *The Morse equivalence $H(A) : X \longrightarrow X_A$ is a homotopy equivalence.*

Furthermore, there is a canonical $A^{(*)}$-grading $\mathrm{gr}_A : X_A^{(*)} \longrightarrow A^{(*)}$ of X_A given by the composition

$$X_A^{(*)} \xrightarrow{\cong} X_{critical}^{(*)} \hookrightarrow X^{(*)} \xrightarrow{\mathrm{gr}_A} A^{(*)}.$$

For all $a \in A^{(*)}$ the subcomplex $(X_A)_{\preceq a}$ with respect to this $A^{(*)}$-grading is given by the subcomplex

$$(X_A)_{\preceq a} = (X_A)_{\preceq_A a}$$

from Definition 4.2.13. In addition, the Morse equivalence $H(A) : X \longrightarrow X_A$ respects these $A^{(*)}$-gradings, that is:

3. *For all $a \in A^{(*)}$ and all $i \in \mathbb{N}$, the i-cells of $(X_A)_{\preceq_A a}$ are in one-to-one correspondence with the A-critical i-cells of $X_{\preceq_A a}$.*
4. *For all $a \in A^{(*)}$ the restriction $H(A)_{\preceq_A a} = H(A)|_{X_{\preceq_A a}}$ is a homotopy equivalence*

$$H(A)_{\preceq_A a} : X_{\preceq_A a} \xrightarrow{\simeq} (X_A)_{\preceq_A a}.$$

Proof. The fact that the A-universal grading $\mathrm{gr}_A : X^{(*)} \longrightarrow A^{(*)}$ of X induces a canonical $A^{(*)}$-grading $\mathrm{gr}_A : X_A^{(*)} \longrightarrow A^{(*)}$ of X_A can be seen from the definition of the Morse complex (see Definition 4.2.13) which is done by induction over the matching poset $A^{(*)}$: Every new cell $\sigma_A \in X_A^{(*)}$ corresponding to a critical cell $\sigma \in X_{critical}^{(*)}$, say $\mathrm{gr}_A(\sigma) = a$ is attached to the complex $(X_A)_{\prec_A a}$ so that for all faces $\tau_A \in X_A$ of σ_A corresponding to critical cells $\tau \in X_{critical}^{(*)}$ we have that $\mathrm{gr}_A(\tau) \prec_A \mathrm{gr}_A(\sigma)$.

The parts *1.* and *3.* are immediate consequences of the inductive construction of the Morse complex: For every element $a \in A^{(*)}$ which is a critical cell

$a = \sigma \in X^{(*)}_{\text{critical}}$, we attach a cell of the same dimension. For every element $a \in A^{(*)}$ which is an element $a = \tau \to \sigma \in A$, we do not attach any cells.

For parts 2. and 4. we use the theory of homotopy colimits. For our combinatorial setting the formulation of the crucial facts given in [62] suits our needs best. From [62, Proposition 3.1] (Projection Lemma) and [62, Proposition 3.7] (Homotopy Lemma) we derive the fact, that the induced maps

$$H(A)_{\preceq_A a} : X_{\preceq_A a} \longrightarrow (X_A)_{\preceq_A a}$$

and

$$H(A) : X \longrightarrow X_A$$

are homotopy equivalences.

Definition 4.2.15 *Let X be a CW-complex, P a partially ordered set, gr : $X^{(*)} \longrightarrow P$ a P-grading of X and A an acyclic matching on X. We call A homogeneous with respect to the P-grading gr : $X^{(*)} \longrightarrow P$ if we have:*

$$\text{gr}(\tau) = \text{gr}(\sigma) \text{ for all } \tau \to \sigma \in A.$$

Remark 4.2.16 *Let X be a CW-complex, P a partially ordered set, gr : $X^{(*)} \longrightarrow P$ a P-grading of X and A an acyclic matching on X. Then the following are equivalent:*

1. *A is homogeneous with respect to the P-grading gr : $X^{(*)} \longrightarrow P$,*
2. *The A-universal grading of X is compatible with the P-grading gr : $X^{(*)} \longrightarrow P$, that is, there exists an order preserving map $g : A^{(*)} \longrightarrow P$ such that the diagram*

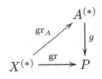

commutes.

This is why we call $\text{gr}_A : X^{(*)} \longrightarrow A^{(*)}$ the universal A-grading of X.

Remark 4.2.17 *Let X be a CW-complex, P a partially ordered set, gr : $X^{(*)} \longrightarrow P$ a compact P-grading of X and A an acyclic matching on X which is homogeneous with respect to gr : $X^{(*)} \longrightarrow P$. Then A is proper.*

Proof. We use again the following commutative diagram:

If $\mathrm{gr} : X^{(*)} \longrightarrow P$ is compact then so is the A-universal grading $\mathrm{gr}_A : X^{(*)} \longrightarrow A^{(*)}$.

The following corollary is the crucial fact that later allows the use of discrete Morse theory for the construction of cellular resolutions in Section 4.3.

Corollary 4.2.18 *Let X be a CW-complex, P a partially ordered set, $\mathrm{gr} : X^{(*)} \longrightarrow P$ a compact P-grading of X and A an acyclic matching on X which is homogeneous with respect to $\mathrm{gr} : X^{(*)} \longrightarrow P$. Then the P-grading $\mathrm{gr} : X^{(*)} \longrightarrow P$ of X induces a compact P-grading $\mathrm{gr} : X_A^{(*)} \longrightarrow P$ of the Morse complex X_A and the Morse equivalence $H(A) : X \longrightarrow X_A$ respects these P-gradings, that is:*

1. *For all $p \in P$ and all $i \in \mathbb{N}$, the i-cells of $(X_A)_{\preceq p}$ are in one-to-one correspondence with the A-critical i-cells of $X_{\preceq p}$.*
2. *For all $p \in P$ the restriction $H(A)_{\preceq p} := H(A)|_{X_{\preceq p}}$ is a homotopy equivalence*
$$H(A)_{\preceq p} : X_{\preceq p} \xrightarrow{\simeq} (X_A)_{\preceq p}.$$

Proof. There is the following commutative diagram:

$$
\begin{array}{ccccc}
 & & & A^{(*)} & \\
 & & \overset{\mathrm{gr}_A}{\nearrow} & \downarrow g & \\
X_A^{(*)} & \xrightarrow{\cong} & X_{\mathrm{critical}}^{(*)} \hookrightarrow & X^{(*)} \xrightarrow{\mathrm{gr}} & P
\end{array}
$$

Here, the composition gr of the maps in the lower row is the induced P-grading of X_A which is order preserving because it factorizes into

$$\mathrm{gr} = g \circ \mathrm{gr}_A$$

where $\mathrm{gr}_A : X_A^{(*)} \longrightarrow A^{(*)}$ is the canonical $A^{(*)}$-grading of X_A.

To prove the remaining assertions, note that

$$X_{\preceq p} = \bigcup_{\substack{a \in A^{(*)}, \\ g(a) \preceq p}} X_{\preceq_A a}$$

and

$$(X_A)_{\preceq p} = \bigcup_{\substack{a \in A^{(*)}, \\ g(a) \preceq p}} (X_A)_{\preceq_A a}.$$

Now we use again [62]: From [62, Proposition 3.1] (Projection Lemma) and [62, Proposition 3.7] (Homotopy Lemma) it follows that the induced maps

$$H(A)_{\preceq p} : X_{\preceq p} \longrightarrow (X_A)_{\preceq p}$$

are homotopy equivalences.

Exercises

Exercise 4.2.3. Describe the matching poset for Example 4.2.2.

Exercise 4.2.4. Set $\Omega_n = \{\{i,j\} \mid 1 \le i < j \le n\}$. Clearly, an element of Ω can be interpreted as an edge in a graph. Thus a subset of Ω_n itself can be seen as a graph. Let Δ_n be the simplicial complex on ground set Ω_n with simplices $B \subseteq \Omega_n$ where B is a disconnected graph on vertex set $[n]$. Using acyclic matchings, show that Δ_n is homotopy equivalent to a wedge of $(n-1)!$ spheres of dimension $n-3$.

Exercise 4.2.5. Let M be a matroid. Show using discrete Morse theory that the simplicial complex M is homotopic to a wedge of spheres of dimension $\mathrm{rk}M - 1$.

Exercise 4.2.6. Show that minimizing the number of critical cells for a discrete Morse function can be formulated as an integer program. (See [28] for a reference and applications of this point of view.)

4.3 Minimizing Cellular Resolutions using Discrete Morse Theory

Now we are in position to formulate the main result of this chapter which makes discrete Morse theory applicable to cellular resolutions.

Theorem 4.3.1 ([3]) *Let $R = k[\Lambda]$ be an affine semigroup ring and M a $\mathbb{Z}^d$-graded R-module. Assume (X, gr) is a compactly $(\mathbb{Z}^d, \Lambda)$-graded CW-complex which supports a cellular resolution $\mathcal{F}_X^{\mathrm{gr}}$ of M. Then for a homogeneous acyclic matching A of X the $(\mathbb{Z}^d, \Lambda)$-graded CW-complex (X_A, gr) supports a cellular resolution $\mathcal{F}_{X_A}^{\mathrm{gr}}$ of M.*

For the proof of Theorem 4.3.1 we need the following lemmas:

Lemma 4.3.2 *Let $\mathcal{C}_X^{\mathrm{gr}}$ be a cellular complex supported by the $(\mathbb{Z}^d, \Lambda)$-graded CW-complex X. Then, for $i \ge 0$ and $\alpha \in \mathbb{Z}^d$*

$$H_i(\mathcal{C}_X^{\mathrm{gr}})_\alpha \cong H_i(X_{\preceq\alpha}; k),$$

as k-vectorspaces.

Proof. For a fixed α an i-cell $c \in X^{(*)}$ gives rise to an α-homogeneous piece $\underline{x}^{\alpha-\gamma}e_c$, $\gamma = \mathrm{gr}(c)$, in homological degree i if and only if $\mathrm{gr}(c) = \gamma \preceq \alpha$. Therefore, as a sequence of k-vectorspaces, $\mathcal{C}_\alpha$ is isomorphic to the cell-homology chain complex of $X_{\preceq\alpha}$, which proves the assertion.

Proposition 4.3.3 *Let C_X^{gr} be the cellular complex supported by the compactly $(\mathbb{Z}^d, \Lambda)$-graded CW-complex X. Let A be a homogeneous acyclic matching on X. Then (X_A, gr) supports a cellular $\mathbb{Z}^d$-graded chain complex $C_{X_A}^{\mathrm{gr}}$ such that $H_i(C_{X_A}^{\mathrm{gr}}) \cong H_i(C_X^{\mathrm{gr}})$, $i \geq 0$.*

Proof. By Lemma 4.3.2 we have

$$H_i(C_{X_A}^{\mathrm{gr}})_\alpha \cong H_i((X_A)_{\preceq\alpha}; k), \ H_i(C_X^{\mathrm{gr}})_\alpha \cong H_i(X_{\preceq\alpha}; k).$$

By Corollary 4.2.18 $H_i(X_{\preceq\alpha}; k) \cong H_i((X_A)_{\preceq\alpha}; k)$ and the assertion follows.

The proof of Theorem 4.3.1 now follows easily.

Proof. Since $\mathcal{F}_X^{\mathrm{gr}}$ is a resolution of M we have $H_i(C_X^{\mathrm{gr}}) = 0$ for $i \geq 1$ and $H_0(C_X^{\mathrm{gr}}) = M$. Now by Proposition 4.3.3 X_A defines a cellular chain complex with $H_i(C_{X_A}^{\mathrm{gr}}) = H_i(C_X^{\mathrm{gr}}) = 0$, $i \geq 1$ and $H_0(C_{X_A}^{\mathrm{gr}}) = H_0(C_X^{\mathrm{gr}}) = M$. Thus X_A and gr support a $\mathbb{Z}^d$-graded free resolution of M as an R-module.

Questions and Exercises

Exercise 4.3.1. Let I be a monomial ideal. Using discrete Morse theory show that the Lyubeznik resolution can be constructed by reducing the Taylor resolution.

Exercise 4.3.2. Let I be a generic monomial ideal. Using discrete Morse theory show that the Scarf resolution can be constructed by reducing the Taylor resolution.

Exercise 4.3.3. Construct the minimal free resolution of k over $k[x_1, \ldots, x_d]$ from the Bar-resolution using discrete Morse theory.

Question 4.3.4. Is it always possible to construct the Hull resolution of a monomial ideal (co-Artinian monomial module) from the Taylor resolution using discrete Morse theory ?

4.4 The Morse Differential

For a finite CW-complex, Forman presents in [20], Chapter 8, a differential complex

$$\mathcal{M} = \xrightarrow{\tilde{\partial}_{p+1}} \ldots \mathcal{M}_p \xrightarrow{\tilde{\partial}_p} \mathcal{M}_{p-1} \longrightarrow \ldots \longrightarrow \mathcal{M}_0$$

with the following properties:

1. For all $p \in \mathbb{N}$ the module $\mathcal{M}_p$ is the free $\mathbb{Z}$-module generated by the critical p-cells of X.
2. The homology of $\mathcal{M}$ is the same as the homology of the underlying CW-complex X.

In [20], Theorem 8.10, he gives an explicit formula for the differential of this complex. In this section we confirm that, for the corresponding acyclic matching A, the differential complex presented by Forman in fact coincides with the cellular chain complex corresponding to the Morse complex X_A. This is an important fact that enables us to present explicit formulas for the differential of resolutions derived via discrete Morse theory for cellular resolutions (see for example [3]).

Definition 4.4.1 *Let X be a CW-complex, A a proper acyclic matching on X and X_A the corresponding Morse complex.*

1. *We denote by*

$$\mathcal{C}^A = \ldots \xrightarrow{\partial^A_{i+1}} \mathcal{C}_i(X_A) \xrightarrow{\partial^A_i} \mathcal{C}_{i-1}(X_A) \xrightarrow{\partial^A_{i-1}} \ldots \xrightarrow{\partial^A_1} \mathcal{C}_0(X_A)$$

the cellular chain complex of X_A and by $[\ :\ ]_A$ the corresponding coefficients of the differential ∂^A, that is, for $\tau \in X_A^{(*)}$ we have

$$\partial^A(\tau) = \sum_{\sigma \text{ facet of } \tau} [\sigma : \tau]_A \, \sigma.$$

2. *For $\sigma \in X^{(*)}$, $\sigma' \in X^{(*)}_{critical}$, we define*

$$\Gamma_A(\sigma, \sigma') := \{\gamma \mid \gamma \text{ is a path in } G^A_X \text{ from } \sigma \text{ to } \sigma'\}.$$

3. *For $\sigma \in X^{(*)}$, $\sigma' \in X^{(*)}_{critical}$, $\dim \sigma = \dim \sigma' = i$,*

$$\gamma = \sigma_0 \to \tau_1 \to \sigma_1 \to \ldots \to \sigma_{k-1} \to \tau_k \to \sigma_k \in \Gamma_A(\sigma, \sigma'),$$

(that is, $\sigma_0 = \sigma$, $\sigma_k = \sigma'$, $\dim \sigma_j = i$, $\dim \tau_j = i+1$ and $\tau_j \to \sigma_{j-1} \in A$ for all $j = 1, \ldots, k$) , we set

$$m(\gamma) := (-1)^k \prod_{i=1}^{k} [\sigma_{i-1} : \tau_i] \cdot [\sigma_i : \tau_i].$$

Theorem 4.4.2 *Let X be a CW-complex, A a proper acyclic matching on X and X_A the corresponding Morse complex. Let $\sigma, \tau \in X^{(*)}_{critical}$, $\sigma_A, \tau_A \in X_A^{(*)}$ their corresponding cells in $X_A^{(*)}$. Then*

$$[\sigma_A : \tau_A]_A = \sum_{\sigma' \text{ facet of } \tau} [\sigma' : \tau] \sum_{\gamma \in \Gamma_A(\sigma', \sigma)} m(\gamma).$$

For the proof of Theorem 4.4.2 we need the following lemmas:

Lemma 4.4.3 *Let X be a CW-complex, $\sigma_0, \tau_0 \in X^{(*)}$, $A = \{\tau_0 \to \sigma_0\}$ a proper (acyclic) matching on X and X_A the corresponding Morse complex. Let $\sigma, \tau \in X^{(*)}_{\text{critical}}$, $\sigma_A, \tau_A \in X^{(*)}_A$ their corresponding cells in X_A. If σ is not a facet of τ_0 or σ_0 is not a facet of τ, then*

$$[\sigma_A : \tau_A]_A = [\sigma : \tau].$$

If σ is a facet of τ_0 and σ_0 is a facet of τ, then

$$[\sigma_A : \tau_A]_A = [\sigma : \tau] - [\sigma_0 : \tau] \cdot [\sigma : \tau_0].$$

Proof. Note that, from Definition 4.2.13, for all critical cells $\sigma \in X^{(*)}_{\text{critical}}$, the restriction of the Morse equivalence $H(A)$ to σ is a homeomorphism onto the cell σ_A in the Morse complex X_A:

$$H(A)|_\sigma : \sigma \xrightarrow{\cong} \sigma_A.$$

These homeomorphisms induce an inclusion

$$i_A^j : X_A^j \hookrightarrow X^j$$

of the j-skeletons. For every j-cell $\sigma \in X^{(j)}$, we get the following commutative diagram, where $p_\sigma : X_A^j \longrightarrow S^j$ is the cellular projection map for the cell σ (see Definition 2.3.1):

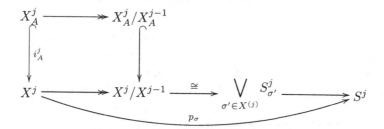

Note that, although i_A^j is not continuous in general, the map

$$X_A^j / X_A^{j-1} \hookrightarrow X^j / X^{j-1}$$

is. For $\sigma_A \in X_A^{(j)}$, we define $p_{\sigma_A} : X_A^j \longrightarrow S^j$ by

$$p_{\sigma_A} := p_\sigma \circ i_A^j.$$

For $\sigma, \sigma' \in X^{(*)}_{\text{critical}}$, we get the commutative diagram from Fig. 4.2.

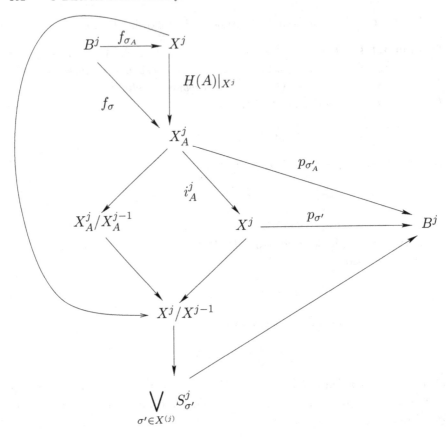

Fig. 4.2. Proof of Lemma 4.4.3, first step

Here, f_σ and f_{σ_A} are the characteristic maps for the cells σ and σ_A. Note that, although this diagram cannot be extended commutatively by adding the identity map id : $X^j \longrightarrow X^j$, it is commutative with both maps $X^j \longrightarrow X^j/X^{j-1}$ defined as the canonical projection maps. We therefore get

$$p_{\sigma'_A} \circ f_{\sigma_A} = p_{\sigma'} \circ f_\sigma.$$

This proves that the p_{σ_A} define cellular projection maps for the cells $\sigma_A \in X_A^{(*)}$.

Now we are ready to calculate, for critical cells $\sigma, \tau \in X_{\text{critical}}^{(*)}$, $\dim \sigma = j$, $\dim \tau = j+1$, the coefficients $[\sigma_A : \tau_A]_A$. For this, we consider the commutative diagram Fig. 4.3.

Using this diagram, straightforward considerations show that if σ is not a facet of τ_0 or σ_0 is not a facet of τ, then we have

$$p_{\sigma_A} \circ f_{\partial \tau_A} = p_\sigma \circ f_{\partial \tau}$$

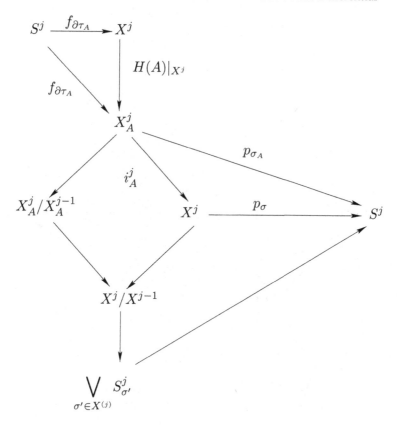

Fig. 4.3. Proof of Lemma 4.4.3, second step

and therefore

$$[\sigma_A : \tau_A] = [\sigma : \tau].$$

Let now $\sigma, \tau \in X_{\text{critical}}^{(*)}$, $\dim \sigma = j$, $\dim \tau = j+1$, such that σ is a facet of τ_0 and σ_0 is a facet of τ. We get the commutative diagram Fig. 4.4.

Here, Ψ is defined to be the unique map such that the diagram Fig. 4.5 commutes. The fact that σ_0 is a regular face of τ_0 shows that

$$\deg\Psi = -[\sigma_0 : \tau_0] \cdot [\sigma : \tau_0].$$

It follows

$$[\sigma_A : \tau_A]_A = [\sigma : \tau] - [\sigma_0 : \tau] \cdot [\sigma : \tau_0].$$

Lemma 4.4.4 *Let X be a CW-complex, $\sigma, \tau \in X^{(*)}$, A an acyclic matching on X, $\tau \to \sigma \in A$ such that $\tau \to \sigma$ is $\prec_A$-minimal in A. Let $A' := A - \{\tau \to \sigma\}$, X_A and $X_{A'}$ the corresponding Morse complexes. Then, for the*

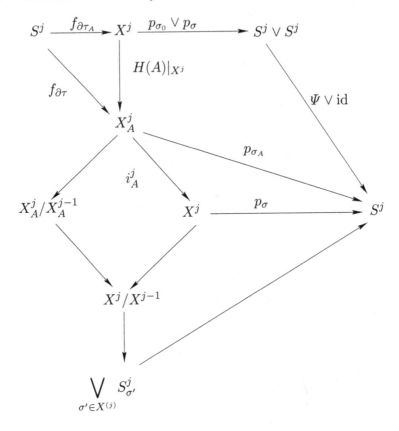

Fig. 4.4. Proof of Lemma 4.4.3, third step

corresponding cells $\sigma_{A'}$, $\tau_{A'} \in X_{A'}^{(*)}$, $\sigma_{A'}$ is a regular facet of $\tau_{A'}$, thus $\{\tau_{A'} \to \sigma_{A'}\}$ is an acyclic matching on $X_{A'}$, and for the corresponding Morse complex $(X_{A'})_{\tau_{A'} \to \sigma_{A'}}$ we have

$$(X_{A'})_{\tau_{A'} \to \sigma_{A'}} = X_A.$$

Furthermore, for the Morse equivalences

$$H(A) : X \xrightarrow{\simeq} X_A,$$
$$H(A') : X \xrightarrow{\simeq} X_{A'}$$

and

$$H(\tau_{A'} \to \sigma_{A'}) : X_{A'} \xrightarrow{\simeq} (X_{A'})_{\tau_{A'} \to \sigma_{A'}},$$

we get the following commutative diagram:

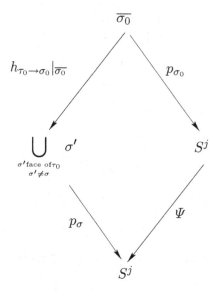

Fig. 4.5. Commutative diagram defining the map Ψ

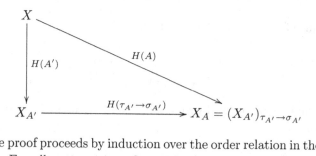

Proof. The proof proceeds by induction over the order relation in the matching poset $A^{(*)}$. For all $a \preceq_A \tau \to \sigma$ the assertion is trivial. For $\tau \to \sigma \preceq a$ the assertion follows from the definition of the Morse complex (see Definition 4.2.13).

After having formulated and proved these facts, we are now ready for the proof of Theorem 4.4.2.

Proof. We can assume that $X_{\text{critical}}^{(*)}$ is finite, since for the coefficient $[\sigma : \tau]$ we only need to consider those $X_{\preceq_A \tau}$ which are compact. We proceed by induction: Let $\sigma_0, \tau_0 \in X^{(*)}$, $\tau_0 \to \sigma_0 \in A$ such that $\tau_0 \to \sigma_0$ is $\prec_A$-minimal in A. Let $A' := A - \{\tau_0 \to \sigma_0\}$, X_A and $X_{A'}$ the corresponding Morse complexes. Then, by induction hypothesis, we have

$$[\sigma_{A'} : \tau_{A'}]_{A'} = \sum_{\sigma' \text{ facet of } \tau} [\sigma' : \tau] \sum_{\gamma \in \Gamma_{A'}(\sigma', \sigma)} m(\gamma),$$

where $\Gamma_{A'}(\sigma',\sigma)$ is the set of paths in $G_X^{A'}$ from σ' to σ.

Note that the following conditions are equivalent:

- $\Gamma_{A'}(\sigma',\sigma_0) = \emptyset$ for all facets σ' of τ
- $\sigma_{0\,A'}$ is not a facet of $\tau_{A'}$ or $\sigma_{A'}$ is not a facet of $\tau_{0\,A'}$.

We distinguish the following cases:

1. The face $\sigma_{0\,A'}$ is not a facet of $\tau_{A'}$ or $\sigma_{A'}$ is not a facet of $\tau_{0\,A'}$. Then

$$\Gamma_A(\sigma',\sigma) = \Gamma_{A'}(\sigma',\sigma)$$

. In addition by Lemma 4.4.3 it follows that

$$[(\sigma_{A'})_{\tau_{0\,A'}\to\sigma_{0\,A'}} : (\tau_{A'})_{\tau_{0\,A'}\to\sigma_{0\,A'}}]_{\tau_{0\,A'}\to\sigma_{0\,A'}} = [\sigma_{A'} : \tau_{A'}]_{A'},$$

. Thus an application of Lemma 4.4.4 yields the asserted formula for $[\sigma_A : \tau_A]_A$.

2. The face $\sigma_{0\,A'}$ is a facet of $\tau_{A'}$ and $\sigma_{A'}$ is a facet of $\tau_{0\,A'}$. Then we have

$$\Gamma_A(\sigma',\sigma) = \Gamma_{A'}(\sigma',\sigma) \cup \{\gamma * (\sigma_0 \to \tau_0 \to \sigma) \mid \gamma \in \Gamma(\sigma',\sigma_0)\}.$$

We deduce from Lemma 4.4.3 that

$$[(\sigma_{A'})_{\tau_{0\,A'}\to\sigma_{0\,A'}} : (\tau_{A'})_{\tau_{0\,A'}\to\sigma_{0\,A'}}]_{\tau_{0\,A'}\to\sigma_{0\,A'}} =$$
$$= [\sigma_{A'} : \tau_{A'}]_{A'} - [\sigma_{0\,A'} : \tau_{A'}]_{A'} \cdot [\sigma_{A'} : \tau_{0\,A'}]_{A'}.$$

The formula now follows by an application of Lemma 4.4.4 and the definition of $m(\gamma)$.

Now we are in position to give a criterion for minimality.

Corollary 4.4.5 *Let (X,gr) be a compactly $(\mathbb{Z},\Lambda)$-graded CW-complex which supports a cellular resolution $\mathcal{F}_X^{\mathrm{gr}}$. Let A be a homogeneous acyclic matching on X. Then $\mathcal{F}_{X_A}^{\mathrm{gr}}$ is minimal if there is no triple of cells $\sigma' \leq \sigma$ and σ'' of X such that σ, σ'' are A-critical, $\mathrm{gr}(\sigma) = \mathrm{gr}(\sigma'')$ and $\dim\sigma' = \dim\sigma'' = \dim\sigma-1$ for which there is a gradient path $\gamma \in \Gamma_A(\sigma',\sigma'')$. In particular, the resolution is minimal if for all A-critical cells σ and $\sigma' \leq \sigma$, $\dim\sigma' = \dim\sigma - 1$ we have $\mathrm{gr}(\sigma') \neq \mathrm{gr}(\sigma)$.*

Proof. The first claim can be deduced from Theorem 4.4.2 using Proposition 3.1.2. The second part of the corollary then is a consequence of the first using the the fact that along gradient paths gr is weakly decreasing.

Exercises

Exercise 4.4.1. Find the minimal number of edges needed in an acyclic matching to construct, starting from the simplex, a CW-complex such that there are coefficients $\neq \pm 1, 0$ in the differential of the Morse complex.

Exercise 4.4.2. Show that it is possible to obtain arbitrarily high coefficients in the differential of the Morse complex of a simplex of sufficiently high dimension.

Question 4.4.3. Let $f(n)$ be the largest absolute value of a coefficient that appears in the Morse differential of an acyclic matching on the n-simplex. What is the asymptotics of $f(n)$?

Exercise 4.4.4. Prove using the construction from Exercise 4.3.2 that in the Scarf resolution all coefficients are ± 1.

References

1. Anderson, I.: Combinatorics of finite sets. Corrected reprint of the 1989 edition. Dover Publications, Mineola, 2002.
2. Athanasiadis, C.A.: Ehrhart polynomials, simplicial polytopes, magic squares and a conjecture of Stanley. J. Reine Angew. Math. 583 (2005) 163-174.
3. Batzies, E., Welker, V.: Discrete Morse theory for cellular resolutions. J. Reine Angew. Math. 543 (2002), 147-168.
4. Batzies, E., Welker, V.: New examples of cellular and minimal free resolutions, in preparation.
5. Bayer, D.,Peeva, I.,Sturmfels, B.: Monomial resolutions. Math. Res. Lett. 5 (1998), 31-46.
6. Bayer, D. , Sturmfels; B.: Cellular resolutions of monomial modules, J. Reine Angew. Math. 502 (1998), 123-140.
7. Bayer, M.M.: Flag vectors of multiplicial polytopes. Electron. J. Comb. 11 (2004) Research paper R65, 13 p.
8. Billera, L.J.,Lee, C.W.: A proof of the sufficiency of McMullen's conditions for f-vectors of simplicial convex polytopes. J. Comb. Theory, Ser. A 31 (1981) 237-255.
9. Björner, A.: Topological methods. Handbook of combinatorics, Vol. 1, 2, 1819–1872, Elsevier, Amsterdam, 1995.
10. Björner, A.,Ziegler, G.M.: Broken circuit complexes: factorizations and generalizations. J. Comb. Theory, Ser. B 51 (1991), 96–126.
11. Bredon, G.E.: Topology and geometry. Graduate Texts in Mathematics **139**. Springer-Verlag, New York, 1993.
12. Bruns, W. , Herzog, J.: Cohen-Macaulay Rings. Combridge University Press, Cambridge, 1993.
13. Bruns, W., Römer, T.: h-vectors of Gorenstein polytopes. J. Comb. Theory, Ser. A 114 (2007) 65-76.
14. Chari, M. K.: On discrete Morse functions and combinatorial decompositions. Formal power series and algebraic combinatorics (Vienna, 1997). Discrete Math. 217 (2000), 101–113.

15. Chari, Manoj K.; Joswig, Michael: Complexes of discrete Morse functions. Discrete Math. 302 (2005) 39-51.

16. Conca, A.,, Hosten, S., Thomas, R.: Nice initial complexes of some classical ideals, math.AC/0512283, to appear in C.A. Athanasiadis, V.V. Batyrev, D.I. Dais, M. Henk, F. Santos Eds., Algebraic and Geometric Combinatorics, Contemporary Mathematics **323**, American Mathematical Society, Providence, 2007.

17. Eagon, J., Reiner, V.: Resolutions of Stanley-Reisner rings and Alexander duality. J. Pure Appl. Algebra 130 (1998) 265–275.

18. Eisenbud, D. : Commutative Algebra with a View Toward Algebraic Geometry. Springer-Verlag, Graduate Texts in Mathematics **150**, New York, 1996.

19. Floystad, G.: Enriched homology and cohomology modules of simplicial complexes. arXiv:math.CO/0411570 J. Alg. Comb. to appear, preprint 2004.

20. Forman, R.: Morse theory for cell complexes. Adv. Math. 134 (1998), 90–145.

21. Forman, R.: Combinatorial differential topology and geometry. New perspectives in algebraic combinatorics (Berkeley, CA, 1996–97), 177–206, Math. Sci. Res. Inst. Publ., 38, Cambridge Univ. Press, Cambridge, 1999.

22. Fröberg, R.: An introduction to Gröbner bases. John Wiley & Sons, Chichester, 1997.

23. Hatcher, A. : Algebraic topology. Cambridge University Press, Cambridge, 2002.

24. Hersh, P., Welker, V.: Gröbner basis degree bounds on $\mathrm{Tor}_\bullet^{k[\Lambda]}(k,k)_\bullet$ and discrete Morse theory for posets. In Barvinok, A., Beck, M., Haase, C., Reznick, B., Welker, V. Eds, Integer points in polyhedra. Geometry, number theory, algebra, optimization. pp. 101-138, Contemporary Mathematics **374**. American Mathematical Society, Providence, 2005.

25. Hibi, T.: Flawless O-sequences and Hilbert functions of Cohen-Macaulay integral domains. J. Pure Appl. Algebra 60 (1989) 245-251.

26. Hibi, T.: Algebraic combinatorics on convex polytopes. Carslaw Publications, Gelbe, 1992.

27. Jöllenbeck, M., Welker, V.: Resolution of the residue class field via algebraic discrete Morse theory, math.AC/0501179, Memoirs Amer. Math. Soc. to appear, preprint 2005.

28. Joswig, M., Pfetsch, M.: Computing optimal discrete morse functions. SIAM J. Discrete Math. 20 (2006), 11-25.

29. Jonsson, J.: Simplicial Complexes of Graphs. PhD-Thesis, KTH-Stockholm, 2005.

30. Jonsson, J. , Welker, V.: A spherical initial ideal for Pfaffians, math.CO/0601335, preprint 2005.

31. Knudson, K., Mramor, N.: Finding critical points discretely, preprint 2006.

32. Kozlov, D.: Discrete Morse theory for free chain complexes. C.R. Math. Acad. Sci. Paris 340 (2005), 867-872.

33. Lyubeznik, G.: A new explicit finite free resolution of ideals generated by monomials in an R-sequence. J. Pure Appl. Algebra 51 (1988), 193–195.

34. McMullen, P.: On simple polytopes. Invent. Math. 113 (1993) 419-444.

35. Miller, E.; Sturmfels, B.: Combinatorial Commutative Algebra, Graduate Texts in Mathematics **227**, Springer-Verlag, New York, 2005.

36. Munkres, J.R.: Topological results in combinatorics. Michigan Math. J. 31 (1984), 113–128.

37. Munkres, J.R.: Elements of algebraic topology. Addison-Wesley Publishing Company, Menlo Park, CA, 1984.

38. Harima, T., Nagel, U., Migliore, J.C., Watanabe, J.: The Weak and Strong Lefschetz properties for Artinian K-Algebras. J. Algebra 262 (2003), 99–126.

39. Nevo, E.: Rigidity and the lower bound theorem for doubly Cohen-Macaulay complexes, math.CO/0505334, preprint 2005, Disc. Comp. Geom. to appear.

40. Niesi, G., Robiano, L.: Disproving Hibi's conjecture with CoCoA; or: Projective curves with bad Hilbert functions. In Eyssette, F., Galligo, A. Eds., Computational algebraic geometry. pp. 195-201, Prog. Math. **109**, Birkhäuser, Boston, 1993.

41. Novik, I.: Lyubeznik's resolution and rooted complexes. J. Algebraic Combin. 16 (2002), 97–101.

42. Novik, I., Postnikov, A., Sturmfels, B.: Syzygies of oriented matroids. Duke Math. J. 111 (2002) 287-317.

43. Ohsugi, H., Hibi, T.: Special simplices and Gorenstein toric rings. J. Comb. Theory, Ser. A 113 (2006) 718-725.

44. Orlik, P., Terao, H.; Arrangements of hyperplanes. Grundlehren der Mathematischen Wissenschaften 300. Springer-Verlag, Berlin, 1992.

45. Pfeifle, J., Ziegler, G. M.: Many triangulated 3-spheres. Math. Ann. 330 (2004) 829-837.

46. Postnikov, A. , Shapiro, B. : Trees, parking functions, syzygies and deformations of monomial ideals. Trans. Amer. Math. Soc. 356 (2004), 3109-3142.

47. Reiner, V., Welker, V.: Linear syzygies of Stanley-Reisner ideals. Math. Scand. 89 (2001), 117-132.

48. Reiner, V. , Welker, V.: On the Charney-Davis and Neggers-Stanley conjectures. J. Comb. Theory, Ser. A 109 (2005) 247-280.

49. Reisner, Gerald: Cohen-Macaulay quotients of polynomial rings. Adv. Math. 21 (1976) 30–48.

50. Römer, T.: Bounds for Betti numbers. J. Algebra 249 (2002) 20-37.

51. Sköldberg, E.: Combinatorial discrete Morse theory from an algebraic viewpoint, Trans. Amer. Math. Soc. 358 (2006) 115-129.

52. Sinefakopoulos, A.. On Borel fixed ideals generated in one degree. In preparstion. 2006.

53. Soll, D. : Polytopale Konstruktionen in der Algebra. PhD-Thesis, Philipps-Universität Marburg, 2006.

54. Soll, D. ,Welker, V. : Type-B generalized triangulations and determinantal ideals. math.CO/0607159, preprint 2006.

55. Stanley, R. P.: Hilbert functions of graded algebra. Adv. Math. 28 (1978) 57-83.

56. Stanley, R. P.: The number of faces of a simplicial convex polytope. Adv. Math. 35 (1980) 236-238.

57. Stanley, R. P.: Combinatorics and commutative algebra. Second edition. Progress in Mathematics, 41. Birkhäuser Boston, Inc., Boston, MA, 1996

58. Schenck, H. K.; Suciu, A. I.: Lower central series and free resolutions of hyperplane arrangements. Trans. Amer. Math. Soc. 354 (2002), 3409–3433.

59. Taylor, D.: Ideals generated by monomials and an R-sequence. Phd Thesis. University of Chicago, 1966.

60. Velasco, M.: Cellular resolutions and nearly Scarf ideals, preprint 2005.

61. Welsh, D.: Matroid Theory. L.M.S. Monographs. 8. Academic Press, London - New York - San Francisco, 1976.

62. Welker, V., Ziegler, G. M.,Živaljević, R. T.: ,omotopy colimits—comparison lemmas for combinatorial applications. J. Reine Angew. Math. 509 (1999), 117–149.

63. Ziegler, G. M., Lectures on polytopes. Graduate Texts in Mathematics **152**. Springer-Verlag, Boston, 1995.

64. Ziegler, G. M.; Živaljević, R. T.: Homotopy types of subspace arrangements via diagrams of spaces. Math. Ann. 295 (1993), 527–548.